Annals of Mathematics Studies

Number 55

ANNALS OF MATHEMATICS STUDIES

Edited by Robert C. Gunning, John C. Moore, and Marston Morse

1. Algebraic Theory of Numbers, *by* HERMANN WEYL
3. Consistency of the Continuum Hypothesis, *by* KURT GÖDEL
11. Introduction to Nonlinear Mechanics, *by* N. KRYLOFF *and* N. BOGOLIUBOFF
20. Contributions to the Theory of Nonlinear Oscillations, Vol. I, *edited by* S. LEFSCHETZ
21. Functional Operators, Vol. 1, *by* JOHN VON NEUMANN
24. Contributions to the Theory of Games, Vol. I, *edited by* H. W. KUHN *and* A. W. TUCKER
25. Contributions to Fourier Analysis, *edited by* A. ZYGMUND, W. TRANSUE, M. MORSE, A. P. CALDERON, *and* S. BOCHNER
27. Isoperimetric Inequalities in Mathematical Physics, *by* G. PÓLYA *and* G. SZEGÖ
28. Contributions to the Theory of Games, Vol. II, *edited by* H. W. KUHN *and* A. W. TUCKER
30. Contributions to the Theory of Riemann Surfaces, *edited by* L. AHLFORS *et al.*
33. Contributions to the Theory of Partial Differential Equations, *edited by* L. BERS, S. BOCHNER, *and* F. JOHN
34. Automata Studies, *edited by* C. E. SHANNON *and* J. MCCARTHY
38. Linear Inequalities and Related Systems, *edited by* H. W. KUHN *and* A. W. TUCKER
39. Contributions to the Theory of Games, Vol. III, *edited by* M. DRESHER, A. W. TUCKER *and* P. WOLFE
40. Contributions to the Theory of Games, Vol. IV, *edited by* R. DUNCAN LUCE *and* A. W. TUCKER
41. Contributions to the Theory of Nonlinear Oscillations, Vol. IV, *edited by* S. LEFSCHETZ
42. Lectures on Fourier Integrals, *by* S. BOCHNER
43. Ramification Theoretic Methods in Algebraic Geometry, *by* S. ABHYANKAR
44. Stationary Processes and Prediction Theory, *by* H. FURSTENBERG
45. Contributions to the Theory of Nonlinear Oscillations, Vol. V, *edited by* L. CESARI, J. LASALLE, *and* S. LEFSCHETZ
46. Seminar on Transformation Groups, *by* A. BOREL *et al.*
47. Theory of Formal Systems, *by* R. SMULLYAN
48. Lectures on Modular Forms, *by* R. C. GUNNING
49. Composition Methods in Homotopy Groups of Spheres, *by* H. TODA
50. Cohomology Operations, *lectures by* N. E. STEENROD, *written and revised by* D. B. A. EPSTEIN
51. Morse Theory, *by* J. W. MILNOR
52. Advances in Game Theory, *edited by* M. DRESHER, L. SHAPLEY, *and* A. W. TUCKER
53. Flows on Homogeneous Spaces, *by* L. AUSLANDER, L. GREEN, F. HAHN, *et al.*
54. Elementary Differential Topology, *by* J. R. MUNKRES
55. Degrees of Unsolvability, *by* G. E. SACKS
56. Knot Groups, *by* L. P. NEUWIRTH
57. Seminar on the Atiyah-Singer Index Theorem, *by* R. S. PALAIS
58. Continuous Model Theory, *by* C. C. CHANG *and* H. J. KEISLER
59. Lectures on Curves on an Algebraic Surface, *by* DAVID MUMFORD
60. Topology Seminar, Wisconsin, 1965, *edited by* R. H. BING *and* R. J. BEAN

DEGREES OF UNSOLVABILITY

BY

Gerald E. Sacks

SECOND EDITION

PRINCETON, NEW JERSEY
PRINCETON UNIVERSITY PRESS
1966

L. C. Card 63-9996

SECOND EDITION 1966
REPRINTED 1993

Printed in the United States of America

Published by Princeton University Press, 41 William Street,
Princeton, New Jersey 08540
In the United Kingdom: Princeton University Press,
Chichester, West Sussex

Printed in the United States of America

Princeton University Press books are printed on acid-free paper
and meet the guidelines for permanence and durability of the
Committee on Production Guidelines for Book Longevity of the Council on
Library Resources

This Book Is For

CLIFFORD SPECTOR, 1930-1961

Ἡδὺ ἡ φίλου μνήμη τεθνηκότος

EPICURUS

PREFACE TO REVISED EDITION

It is intended that Section 1 be read before any section that follows it; the same applies to Sections 4 and 8. There is a continuous commentary on the priority method which starts in Section 4 and ends in Section 10.

We wish to thank the many persons who helped up prepare and revise this monograph at one stage or another. Among them were R. Friedberg, S. C. Kleene, A. Nerode, A. Robinson, H. Rogers, Jr., J. Rosenstein, J. R. Shoenfield, R. Smullyan, and C. Spector. Special thanks are owed to J. Barkley Rosser. We also thank the National Science Foundation and the Army Research Office (Durham) for financial support.

Many results on degrees were obtained after the appearance of the first printing of this monograph. We discuss some of them in Section 12. Nonetheless, we still feel that the subject of degrees is far from finished. Many of the weapons developed to attack degrees are now making inspiring appearances on the other battlefields of mathematical logic, particularly set theory. We hope that what follows will hasten the final victory.

Ithaca, New York
April 1, 1966

CONTENTS

§1. Preliminaries 1

§2. A Continuum of Mutually Incomparable Degrees. 7

§3. Uncountable Suborderings of Degrees 21

§4. The Priority Method of Friedberg and Muchnik. 43

§5. An Existence Theorem for Recursively Enumerable Degrees 55

§6. The Jump Operator 77

§7. An Interpolation Theorem for Recursively Enumerable Degrees . . 117

§8. Minimal Upper Bounds for Sequences of Degrees 123

§9. Minimal Degrees 135

§10. Measure-theoretic, Category and Descriptive Set-theoretic Arguments 153

§11. Initial segments of Degrees 163

§12. Further Results and Conjectures 169

BIBLIOGRAPHY. 173

DEGREES OF UNSOLVABILITY

DEGREES OF UNSOLVABILITY

§1. Preliminaries

The natural numbers are 0,1,2,...; let f and g be functions from the natural numbers into the natural numbers. We say f and g have the same degree of recursive unsolvability if f is recursive in g and g is recursive in f. We use degree to mean degree of recursive unsolvability. The degree of f is the set of all functions g such that f is recursive in g and g is recursive in f. We denote the degree of f by $\underline{f}$; in general, we use underlined, lower case, Roman letters to denote degrees. Let U denote the set of all degrees. We give U a partial ordering as follows: $\underline{f} \leq \underline{g}$ if f is recursive in g. We claim U is an upper semi-lattice. Let $\underline{f}$ and $\underline{g}$ be members of U. We define

$$h(n) = 2^{f(n)} \cdot 3^{g(n)}$$

for all n. Then f and g are each recursive in h; in addition, if f and g are each recursive in k, then h is recursive in k. But then $\underline{h}$ is the least member of the set of all $\underline{k}$ such that $\underline{f} \leq \underline{k}$ and $\underline{g} \leq \underline{k}$. We call $\underline{h}$ the least upper bound or union of $\underline{f}$ and $\underline{g}$, or in symbols, $\underline{h} = \underline{f} \cup \underline{g}$. Saying that f is recursive in $g_0, g_1, \ldots, g_n$ is the same as saying $\underline{f} \leq \underline{g}_0 \cup \underline{g}_1 \cup \ldots \cup \underline{g}_n$. A set S of degrees is called independent if no member of S is less than or equal to any finite union of other members of S.

The notion of degree is due to Post [17]. The upper semi-lattice of degrees was first defined and studied by Post and Kleene in [9].

We write $\underline{f}|\underline{g}$ to indicate $\underline{f} \nleq \underline{g}$ and $\underline{g} \nleq \underline{f}$.

We assume a knowledge of elementary recursion theory; our basic source of notation is Kleene [7]. We do, however, extend some of Kleene's notation.

Suppose f is recursive in g. Then there exists a natural number e such that e is the Gödel number of a system of equations which defines f recursively in g. Since the set of natural numbers is countable, there can only be countably many f's recursive in a given g. But then there can only be countably many $\underline{f}$'s such that $\underline{f} \leq \underline{g}$. Thus each degree has at most countably many predecessors. (If $m, n \in P$, where P is partially ordered, we say m is a predecessor of n if $m \leq n$.) Each degree consists of countably many functions. It follows there is a continuum of degrees, since there is a continuum of functions.

The completion of a function f is the representing function of the predicate

$$(Ey)T_1^1\left(\tilde{f}(y),\ (n)_0,\ (n)_1,\ y\right) \quad ;$$

Kleene [7, XI] shows that any predicate of the form (Ey)R(n,y), where R is recursive in f, is recursive in the completion of f. It follows that if f is recursive in g, then the completion of f is recursive in the completion of g. With each degree $\underline{f}$ we associate its jump $\underline{f}'$; $\underline{f}'$ is the degree of completion of f. We have

$$\underline{f} \leq \underline{g} \rightarrow \underline{f}' \leq \underline{g}' \quad .$$

Thus the jump operator is order-preserving. A degree is called complete if it is the jump of some other degree. The jump operator was first defined in [9]. Kleene [7, XI] shows that the completion of f is not recursive in f. Thus

$$\underline{f} < \underline{f}' \quad ,$$

since f is recursive in the completion of f.

We study the ordering of U in order to learn how functions are classified by the notion of relative recursiveness. The jump operator is important since it indicates the role of quantification in the above classification. All the theorems in this monograph will deal solely with the ordering of U. We will not discuss or determine the degrees of particular functions or sets, as Rogers did in [18].

The degree of a set or predicate is the degree of its representing function. For each function f there is a set A which has the same degree as f:

$$2^n \cdot 3^m \in A \longleftrightarrow m = f(n) \quad .$$

In some of our arguments we will assume that all functions are representing functions.

Let $\underline{0}$ be the degree of all recursive functions. Thus $\underline{0} \leq \underline{f}$ for all $\underline{f}$. For each $n > 0$, we define $\underline{d}^{(n)} = \left(\underline{d}^{(n-1)}\right)'$, where $\underline{d}^{(0)} = \underline{d}$. We call a degree arithmetical if it is less then $\underline{0}^{(n)}$ for some n. The arithmetical degrees are just those degrees which are degrees of predicates which are arithmetical in the sense of Kleene [7, IX].

A degree $\underline{d}$ is said to be recursively enumerable in a degree $\underline{c}$ if some set of degree $\underline{d}$ is the range of a function of degree less than or equal to $\underline{c}$. Then $\underline{c}'$ is recursively enumerable in $\underline{c}$, and $\underline{c}'$ is the greatest of all those degrees recursively enumerable in $\underline{c}$; $\underline{d}$ is called recursively enumerable if it is the degree of a recursively enumerable set. The recursively enumerable degrees are just those degrees which are recursively enumerable in $\underline{0}$. The notion of $\underline{d}$ being recursively enumerable in $\underline{c}$ is due to Shoenfield [23].

If f is a function, let f' be its completion. Suppose f is recursive in g with Gödel number e. Then the predicate

$$T_1^1\left(\tilde{f}(y),\ (n)_0,\ (n)_1,\ y\right)$$

is recursive in g with Gödel number e*; also, e* can be found effectively if e is known. But then there is a Gödel number e** such that

$$(Ey)T_1^1\left(\tilde{f}(y),\ (n)_0,\ (n)_1,\ y\right)$$

is recursive in g' with Gödel number e**; also, e** can be found effectively if e* is known (Kleene [7, XI]). Thus there is a recursive function k such that for any f and g, if f is recursive in g with Gödel number e, then f' is recursive in g' with Gödel number k(e).

For each f, let $f^{(0)} = f$, let $f^{(m)} = \left(f^{(m-1)}\right)'$ for each $m > 0$, and let $f^{(\omega)}$ be defined by $f^{(\omega)}(m, n) = f^{(m)}(n)$. Suppose f is recursive in g, with Gödel number d. Let $t(0) = e$, and let $t(m) = k(t(m-1))$ for all $m > 0$. Then $f^{(m)}$ is recursive in $g^{(m)}$ with Gödel number t(m). Since t is recursive, it follows $f^{(\omega)}$ is recursive in

$g^{(\omega)}$. Thus if f and g have the same degree, so do $f^{(\omega)}$ and $g^{(\omega)}$. With each degree $\underline{f}$ we associate its transfinite jump $\underline{f}^{(\omega)}$; $\underline{f}^{(\omega)}$ is the degree of $f^{(\omega)}$. It follows

$$\underline{f}^{(n)} < \underline{f}^{(\omega)}$$

for all n. Thus $\underline{0}^{(\omega)}$ is not arithmetical, but it is an upper bound for all the arithmetical degrees. If $\underline{f}$ is arithmetical, then $\underline{f}^{(\omega)} = \underline{0}^{(\omega)}$. The transfinite jump was defined by Kleene and Post [9]; it corresponds to the infinitary logical notion of regarding the number of quantifiers in a predicate as a variable. The definition of the transfinite jump was suggested by the standard construction [7, XI] of a hyperarithmetic predicate which is not arithmetic.

Let us recall some definitions due to Kleene [7, 8]. Let f be a function. We recall:

$$\tilde{f}(y) = \prod_{i < y} p_i^{f(i)} \quad ;$$

$$\bar{f}(y) = \prod_{i < y} p_i^{1+f(i)} \quad ;$$

$$\text{Seq}\ (x) \leftrightarrow (i)(i < \ell h(x) \rightarrow (x)_i > 0) \quad .$$

We define:

$$Cv(x) = \prod_{i < \ell h(x)} p_i^{(x)_i \dot{-} 1} \quad .$$

Then $Cv(\bar{f}(y)) = \tilde{f}(y)$ for all y. We will often take the liberty of writing

$$T_1^1(\bar{f}(y),\ e,\ n,\ y)$$

when we mean

$$T_1^1(Cv(\bar{f}(y)),\ e,\ n,\ y) \quad .$$

More generally, we will often write $T_1^1(x, e, n, y)$ when we mean $T_1^1(Cv(x), e, n, y)$; context will always make clear what is meant. We read "Seq(x)" as x is a sequence number. We can tell effectively what the length of a sequence number is.

Let h be a partial (possibly total function). We define the partial function $\{e\}^h$ as follows:

$$\{e\}^h(n) \simeq U\Big(\mu y\Big[(i)_{i<y}\big(h(i) \text{ is defined}\big) \;\&\; T_1^1\big(\tilde{h}(y),\, e,\, n,\, y\big)\Big]\Big)$$

If h is total, then our definition of $\{e\}^h$ agrees with that of Kleene [7, XII]. If h is not total, then our definition of $\{e\}^h$ does not necessarily agree with that of other authors. Let s be a sequence number. Let h be a partial function whose domain is the length of s and whose values are given by: $h(i) = (s)_i - 1$ for all $i < \ell h(s)$. We define

$$\{e\}^s(n) \simeq \{e\}^h(n)$$

for all n. If s is a sequence number such that $\{e\}^s(n)$ is defined for a given e and n, then

$$\{e\}^s(n) = \{e\}^h(n)$$

for all h such that $\bar{h}(t) = s$ for some t.

Suppose f is the unique function defined by:

$$f(n) = H\big(\tilde{f}(n),\, n\big) \quad ;$$

then f is recursive in H. Suppose f is the unique function defined by

$$f(n) = \begin{cases} H_1\big(\tilde{f}(n),\, n\big) & \text{if} \quad (Ex)(y)R\big(x,\, y,\, \tilde{f}(n)\big) \\ H_2\big(\tilde{f}(n),\, n\big) & \text{otherwise} . \end{cases}$$

Then f is recursive in H_1, H_2, r'', where r is the representing function of R; in particular, if H_1, H_2 and r are recursive, then $\underline{f} \leq \underline{0}''$. If f is a function defined by induction, we can often obtain an upper bound for the degree of f by examining the induction step of the definition of f. This last will be our method in many of the theorems that follow. We will not, as a rule, completely formalize the definition of f, but we will always write enough equations to justify our claims concerning an upper bound on the degree of f. Several examples of this method are worked out in great detail in [9].

We follow the view that recursion theory is just another branch of mathematics, and that proofs in recursion theory should be presented with no more detail than proofs in algebraic topology or Fourier analysis. This view apparently first appeared in Post [17] in 1944, but was largely ignored until the early 1950's when it reappeared in the papers of Dekker,

Myhill, Friedberg and others. Thus we resort to equations only when they seem to clarify matters. We use words like "effective" and "computable" whenever we feel there is no clear and present danger of being misunderstood.

§2. A Continuum of Mutually Incomparable Degrees

In [22] J. R. Shoenfield proved with the aid of Zorn's lemma that there exists an uncountable set of mutually incomparable degrees. In this section we prove without any use of the axiom of choice that there exists a set of mutually incomparable degrees whose cardinality is that of the continuum. This last result was first obtained by D. Lacombe in an unpublished paper; we give a different and simpler proof below. In this section we also prove (following Spector [25]) that no infinite, strictly ascending sequence of degrees has a least upper bound, and that the upper semi-lattice of degrees less than or equal to $\underline{0}''$ is not a lattice.† All the results of this section are based on techniques occuring in Kleene and Post [9].

> THEOREM 1. Let A, B and C be countable sets of functions with the following properties: if $f \in B \cup C$, then f is not recursive in the members of any finite subset of $A \cup (C - \{f\})$; no member of C is recursive in the members of any finite subset of B. Then there exists a function $g \notin C$ such that if $C' = C \cup \{g\}$, then the above statement remains true when C is replaced by C'; in addition, each member of A is recursive in g.

PROOF. Let $A = \{a_i | \geq 0\}$, $B = \{b_i | i \geq 0\}$ and $C = \{c_i | i \geq 0\}$. The construction of g proceeds by stages. At stage $s \geq 0$, we define finitely or infinitely many values of g. Before we proceed with stage s, we state an induction hypothesis concerning what has taken place in the course of the construction prior to state s:

† Spector [25] also indicates how to show the upper semi-lattice of degrees less than or equal to $\underline{0}'$ is not a lattice.

(H1) there are only finitely many n such that $g(n)$ has been defined and n is not a power of a prime p_j for some $j < s$.

(H2) if $2^j < s$, then there is a $t(j) > 0$ such that $g(p_j^{t(j)+n})$ has been set equal to $a_j(n)$ for all n;

(H3) if $g(p_j^m)$ has been defined for infinitely many m, then $2^j < s$.

One of four cases must be dealt with at stage s.

CASE 1. $s = 2^j$. The object of this case is to have a_j recursive in g when the construction of g is completed. It follows from (H3) that there is a t such that $g(p_j^{t+n})$ has not yet been defined for any $n \geq 0$; let $t(j)$ be the least such positive t. We set $g(p_j^{t(j)+n}) = a_j(n)$ for all n.

CASE 2. $(s)_0 = 0$, $(s)_1 = 0$ and $s \neq 1$. Let $(s)_2 = k$, $(s)_3 = e$ and $(s)_4 = m$. The purpose of this case is to insure that b_k will not be recursive in $a_0, a_1, \ldots, a_m, c_0, c_1, \ldots, c_m, g$ with Gödel number e when the construction of g is completed. Note that $B \cap C = 0$, since no member of C is recursive in the members of any finite subset of B. Before we can dispose of Case 2, we must recall several notions discussed in Section 1. If h is a function, then $\{e\}^h$ is the partial function defined by

$$\{e\}^h(n) \simeq U\Big(\mu y(m)_{m<y}\Big[h(m) \text{ is defined } \& \ T_1^1\big(\tilde{h}(y), e, n, y\big)\Big]\Big) \quad .$$

If h and h^* are partial functions such that the domain of h is strictly contained in the domain of h^* and such that h and h^* agree on the domain of h, then we say h^* extends h. Note that if h^* extends h and $\{e\}^h(n)$ is defined, then $\{e\}^{h^*}(n)$ is defined and is equal to $\{e\}^h(n)$. We say h^* is a finite extension of h if h^* is an extension of h and the domain of h^* differs only finitely from the domain of h. If K is a finite set of partial functions, then we define $\{e\}^K$ in the same fashion as $\{e\}^h$.

Let g_s be the unique partial function with the property that $g_s(n)$ is defined and equal to $g(n)$ if and only if $g(n)$ has been defined prior to state s. If g^* is a finite extension of g_s, we define the weight of g^* to be the product of the members of the set

$$\left\{p_i^{1+g^*(i)} \mid g^*(i) \text{ is defined \& } g_s(i) \text{ is not defined}\right\} \quad ;$$

we define the weight of g_s to be 0. The concept of weight provides us with a convenient well-ordering of the set of all finite extensions of g_s.

CASE 2.1. There exists a finite extension of g^* of g_s such that for some n,

$$\{e\}^{a_0, \ldots, a_m, c_0, \ldots, c_m, g^*}(n)$$

is defined but not equal to $b_k(n)$.. Among all such finite extensions, there is a unique one of least weight. Let it be denoted by g'. We set $g(m) = g'(m)$ for all m for which $g'(m)$ is defined and $g_s(m)$ is undefined. There are, of course, only finitely many such m.

CASE 2.2. No such finite extension exists. We proceed directly to stage $s + 1$.

CASE 3. $(s)_0 = 0$ and $(s)_1 = 1$. Let $(s)_2 = k$, $(s)_3 = e$ and $(s)_4 = m$. If $m \leq k$, we proceed directly to stage $s + 1$. Let $m > k$. We take steps to guarantee that c_k will not be recursive in $a_0, a_1, \ldots, a_m, c_0, c_1, \ldots, c_{k-1}, c_{k+1}, \ldots, c_m, g$ with Gödel number e when the construction of g is completed.

CASE 3.1. There exists a finite extension g^* of g_s such that for some n, $\{e\}^{K, g^*}(n)$ is defined but not equal to $c_k(n)$ where

$$K = \{a_0, a_1, \ldots, a_m, c_0, c_1, \ldots, c_{k-1}, c_{k+1}, \ldots, c_m\} \quad .$$

We proceed as in Case 2.1.

CASE 3.2. Same as Case 2.2.

CASE 4. $(s)_0 = 0$ and $(s)_1 \geq 2$. Let $(s)_2 = e$ and $(s)_3 = m$. The object of this case is to make sure that g will not be recursive in $a_0, a_1, \ldots, a_m, b_0, b_1, \ldots, b_m, c_0, c_1, \ldots, c_m$ with Gödel number e when the construction of g is completed. Let r_s be the least r such that $g(r)$ has not yet been defined. Let

$$J = \{a_0, a_1, \ldots, a_m, b_0, b_1, \ldots, b_m, c_0, c_1, \ldots, c_m\} \quad .$$

If $\{e\}^J(r_s)$ is defined, we set $g(r_s)$ equal to the least non-negative integer different from it. If it is not defined, we set $g(r_s)$ equal to 0.

That finishes the definition of g. The induction hypothesis for stage $s + 1$ is readily verified, since Cases 2, 3, and 4 permit only finite extensions of g_s at stage s. By Case 4, g is defined everywhere. By Case 1 each member of A is recursive in g. By Case 4, g is not recursive in the members of any finite subset of $A \cup B \cup C$. Thus $g \notin C$.

We show that b_k is not recursive in $a_0, a_1, \ldots, a_m, c_0, c_1, \ldots, c_m, g$ with Gödel number e. Let

$$s = 5^k \cdot 7^e \cdot 11^m \cdot 13 \quad .$$

Then Case 2 is dealt with at stage s. Let $W = \{a_0, a_1, \ldots, a_m, c_0, c_1, \ldots, c_m\}$. Suppose Case 2.1 holds at stage s. Then for some n, $\{e\}^{W,g}(n)$ is defined but not equal to $b_k(n)$. Suppose Case 2.2 holds at stage s. There are now two possibilities. First, there may be an n such that $\{e\}^{W,g^*}(n)$ is undefined for every g^* which is a finite extension of g_s. If this last is true, then $\{e\}^{W,g}(n)$ is undefined for some n, and consequently, b_k is not recursive in W, g with Gödel number e. Second, for every n there is a g^* such that g^* is a finite extension of g_s and such that $\{e\}^{W,g^*}(n)$ is defined. We claim that the second possibility cannot happen. Suppose it does. For each n, let $z(n)$ be

$$\mu x(Ey)_{y<x}(Eg^*)\Big(g^* \text{ is a finite extension of } g_s$$
$$\& \ (m)_{m<y}(g^*(m) \text{ is defined}) \ \& \ x \text{ is the weight of } g^*$$
$$\& \ T_1^{W,g^*}(e, n, y)\Big) \ .$$

Let D_s be the domain of g_s. Then z is computable from D_s, g_s, W. It follows from (H1)-(H3) (induction hypothesis for stage s) that the set D_s is recursive and that the values of the partial function g_s can be computed effectively from $a_0, a_1, \ldots, a_{s-1}$. Thus z is recursive in the members of $W \cup \{a_0, a_1, \ldots, a_{s-1}\}$. For each n, let $v(n) = \{e\}^{W,g^*}(n)$, where g^* is the unique, finite extension of g_s of weight $z(n)$. Since Case 2.2 holds at stage s, $v(n) = b_k(n)$ for all n. But then b_k is recursive in the members of a finite subset of $A \cup C$. This last contradicts the hypothesis of Theorem 1.

It remains only to show that c_k is not recursive in $a_0, a_1, \ldots, a_m, c_0, c_1, \ldots, c_{k-1}, c_{k+1}, \ldots, c_m, g$ with Gödel number e.

Let

$$s = 3 \cdot 5^k \cdot 7^e \cdot 11^m \quad .$$

Then Case 3 is dealt with at stage s. Let K be defined as in Case 3.1. If Case 3.1 holds at stage s, then for some n, $\{e\}^{K,g}(n)$ is defined but not equal to $c_k(n)$. Suppose Case 3.1 does not hold. There are, once again, two possibilities. First, there may be an n such that $\{e\}^{K,g^*}(n)$ is undefined for every g^* which is a finite extension of g_s. If this last is true, then $\{e\}^{K,g}(n)$ is undefined for some n, and all is well. Second, for every n there is a g^* such that g^* is a finite extension of g_s and $\{e\}^{K,g^*}(n)$ is defined. We claim that the second possibility cannot happen. Suppose it does. We proceed now as we did above with Case 2. It follows that c_k is recursive in the members of a finite subset of $A \cup (C - \{c_k\})$. But this last contradicts the hypothesis of Theorem 1.

We have assumed that A, B and C are infinite; no substantial change in the argument is required if any of the three sets are finite.

COROLLARY 1. (Spector [25]) Let S be a countable set of degrees. Then either S has no least upper bound or S has a least upper bound which is equal to the least upper bound of a finite subset of S.

PROOF. Let $S = \{\underline{a}_i | i > 0\}$. Suppose S does not have a least upper bound which is equal to the least upper bound of a finite subset of S. Let $\underline{b}$ be an upper bound for S. We show $\underline{b}$ is not a least upper bound for S. Let b be a function of degree $\underline{b}$, and for each i, let a_i be a function of degree $\underline{a}_i$. Let $A = \{a_i | i > 0\}$, $B = \{b\}$ and $C = 0$. By Theorem 1, there is a function g with the following properties; b is not recursive in g, g is not recursive in b, and each member of A is recursive in g. Let $\underline{g}$ be the degree of g. Then $\underline{g}$ is an upper bound for S and $\underline{g} | \underline{b}$.

In [25] Spector left open the question of whether or not any or all infinite, ascending sequences of degrees have minimal upper bounds. In Section 5, we construct an infinite, ascending sequence of degrees which has $\underline{0}'$ as a minimal upper bound. In Section 6, we show that no infinite, ascending sequence of simultaneously recursively enumerable degrees has $\underline{0}'$ as a minimal upper bound. In Section 8, we show that every countable set of degrees has a minimal upper bound!

COROLLARY 2. Let A and B be countable sets of functions such that no member of B is recursive in the members of any finite subset of A. Then there exists a countably infinite set C of functions such that $B \cap C = 0$ and such that if $f \in B \cup C$, then f is not recursive in the members of any finite subset of $A \cup (C - \{f\})$.

PROOF. We proceed by induction on s. Suppose $C_s = \{c_0, c_1, \ldots, c_{s-1}\}$ is a set of functions such that (1) if $f \in B \cup C_s$, then f is not recursive in the members of any finite subset of $A \cup (C_s - \{f\})$, and (2) no member of C_s is recursive in the members of any finite subset of B. If $s = 0$, then $C_s = 0$ and our supposition is merely the hypothesis of Corollary 2. By Theorem 1, there is a function $g \notin C_s$ such that (3) if $f \in B \cup C_{s+1}$, where $C_{s+1} = C_s \cup \{g\}$, then f is not recursive in the members of any finite subset of $A \cup (C_{s+1} - \{f\})$, and such that (4) no member of C_{s+1} is recursive in the members of any finite subset of B. Let $C = \cup \{C_s | s \geq 0\}$. Then $B \cap C = 0$ by (4).

THEOREM 2. If A and B are countable sets of functions, then (1) and (2) are equivalent: (1) no member of B is recursive in the members of any finite subset of A; (2) there is a set D of functions of cardinality of the continuum such that no member of D is recursive in any member of D other than itself, each member of A is recursive in every member of D, and no member of B is recursive in any member of D.

PROOF. We show that (1) entails (2). First we note an immediate corollary of Theorem 1: (N) if F and G are countable sets of functions such that no member of G is recursive in the members of any finite subset of F, then there exists a function t such that every member of F is recursive in t, but no member of G is recursive in t. For each n let $K_n = \{2n, 2n+1\}$. Let W be the set of all functions such that $f \in W$ if and only if $f(n) \in K_n$ for all n. Then W has cardinality of the continuum; also, if f and g are distinct members of W, then the range of f is not contained in the range of g. By Corollary 2 above, there exists a set $C = \{c_0, c_1, c_2, \ldots\}$ of distinct functions such that

$B \cap C = 0$ and such that if $f \in B \cup C$, then f is not recursive in the members of any finite subset of $A \cup (C - \{f\})$. Then for each $g \in W$, we have that no member of

$$B \cup \{c_i \mid i \notin \text{range of } g\}$$

is recursive in the members of any finite subset of $A \cup \{c_i \mid i \in \text{range of } g\}$. It follows from note (N) above that there exists a map t defined on W such that for each $g \in W$, $t(g)$ is a function with the property that (a) every member of $A \cup \{c_i \mid i \in \text{range of } g\}$ is recursive in $t(g)$, but (b) no member of $B \cup \{c_i \mid i \notin \text{range of } g\}$ is recursive in $t(g)$.

Observe that the map t is definable without any use of the axiom of choice, since in the proof of Theorem 1, the function g is defined uniquely from A, B and C without any free act of choice. Choices are made in the course of the proof of Theorem 1, but they are made with the help of the natural well-ordering of the natural numbers and the given well-orderings of A, B and C.

Let $D = \{t(g) \mid g \in W\}$. Then each member of A is recursive in every member of D and no member of B is recursive in any member of D. Suppose f and g are distinct members of W such that $t(f)$ is recursive in $t(g)$. Then each member of $\{c_i \mid i \in \text{range of } f\}$ is recursive in $t(f)$ and hence in $t(g)$. But then it follows from the definition of t that

$$\{c_i \mid i \in \text{range of } f\} \subseteq \{c_i \mid i \in \text{range of } g\}.$$

This last is impossible, since f and g are distinct members of W and the c_i's are distinct. We conclude that D has cardinality of the continuum and that no member of D is recursive in any member of D other than itself.

COROLLARY 1. There exists a set of mutually incomparable degrees of cardinality of the continuum.

The proof of Corollary 1 does not require the axiom of choice.

In [9] Kleene and Post showed that the upper semi-lattice of degrees is not a lattice. In [25] Spector showed that upper semi-lattice of degrees of arithmetical sets is not a lattice; in fact, he proved that the upper semi-lattice of degrees less than or equal to $\underline{0}''$ is not a lattice.

We follow his argument below.

THEOREM 3. The upper semi-lattice of degrees less than or equal to $\underline{0}''$ is not a lattice.

PROOF. Our argument has two parts. First we construct an infinite, ascending sequence of degrees less than $\underline{0}'$. Then we make use of this sequence to define two degrees less than $\underline{0}''$ with no greatest lower bound.

We define a function $\lambda in|A(i, n)$ by stages. At stage s we define values of $A(i, n)$ for finitely many i and n. For each i, let a_i denote $\lambda n|A(i, n)$, and let a^i denote

$$\lambda jn|A\Big(j + sg((j+1) \dot{-} i), n\Big) .$$

At stage s we take steps to insure that a_i will not be recursive in a^i when the construction of $\lambda in|A(i, n)$ is completed. We make use of the notions of finite extension and weight in a manner similar to that of the proof of Theorem 1.

CASE 1. Let $(s)_0 = i$ and $(s)_1 = e$. Let a^i_s be the unique, partial function whose domain consists of all pairs (j, n) such that $a^i(j, n)$ has been defined prior to stage s and whose values are given by $a^i_s(j, n) = a^i(j, n)$. Let $a_{i,s}$ be the unique partial function whose domain consists of all n such that $a_i(n)$ has been defined prior to stage s and whose values are given by $a_{i,s}(n) = a_i(n)$. If b is a partial function of two variables such that the domain of b strictly contains the domain of a^i_s and such that b agrees with a^i_s on the domain of a^i_s, then we say b is an extension of a^i_s. We say b is a finite extension of a^i_s if b is an extension of a^i_s and if the domain of b differs only finitely from the domain of a^i_s. If b is a partial function of two variables, then we define the partial function $\{e\}^b$ by

$$\{e\}^b(n) \simeq U\Big(\mu y(m)_{m<y}(v)_{v<y}\Big[b(m, v) \text{ is defined}$$
$$\& \ T^2_1\Big(\tilde{b}(y, y), e, n, y\Big)\Big]\Big) .$$

If b is a partial function of two variables with a finite domain, we define the weight of b to be the product of the members of the set

$$\left\{p_j^{1+z(j)} \middle| b(j,\ n) \text{ is defined for some } n\right\}$$, where

$z(j)$ is the product of the members of the set $\{p_n^{1+b(j,n)} | b(j,n)$ is defined$\}$. The notion of weight provides us with a convenient well-ordering of the set of all finite extensions of $a_s^{\,i}$. Our purpose at stage s is to choose one or none of these extensions.

Let r_s be the least n for which $a_{i,s}(n)$ is undefined.

CASE 1.1. There exists a finite extension b of a_s^i such that

$$\{e\}^b(r_s)$$

is defined. Among all such extensions there is a unique one of least weight; let it be denoted by b_s. We define

$$a^i(j,\ n) = b_s(j,\ n)$$

for all $(j,\ n)$ such that $b_s(j,\ n)$ is defined and $a_s^i(j,\ n)$ is not defined. We define

$$a_i(r_s) = 1 + \{e\}^{b_s}(r_s) \quad .$$

CASE 1.2. No such finite extension exists. We set $a_i(r_s) = 0$.

That completes the definition of $\lambda in|A(i,\ n)$. At stage s we defined the value of $A(i,\ r_s)$, where $i = (s)_0$ and r_s is the least n such that $A(i,\ n)$ was not defined prior to stage s. This last fact guarantees that $A(i,\ n)$ has been defined for every $(i,\ n)$. To see that a_i is not recursive in a^i with Gödel number e, let $s = 2^i \cdot 3^e$. Suppose Case 1.1 holds at stage s. Then $\{e\}^{a^i}(r_s)$ is defined and equal to $\{e\}^{b_s}(r_s)$, since a^i is an extension of b_s. But then

$$a_i(r_s) = 1 + \{e\}^{a^i}(r_s) \neq \{e\}^{a^i}(r_s) \quad .$$

If Case 1.2 holds, then $\{e\}^{a^i}(r_s)$ is not defined, since a^i is an extension of a_s^i.

Now we show that the degree of $\lambda in|A(i,\ n)$ is less than or equal to $\underline{0}'$. We assume familiarity with the closing remarks of Section 1. We introduce a recursive predicate P:

$P(x,\ y,\ n,\ e,\ d) \longleftrightarrow x$ and y are weights of partial functions of two variables and finite domains

& the partial function whose weight is x extends the one whose weight is y

$\quad\&\ \ b$ is the partial function represented by x

$\quad\&\ \ (m)_{m<d}(v)_{v<d}(b(m, v)$ is defined)

$\quad\&\ \ T_1^2(\tilde{b}(d, d), e, n, d)$.

The last three clauses of $P(x, y, n, e, d)$ say that x is the weight of a finite, partial function b with the property that $\{e\}^b(n)$ is defined. Since P is recursive, the predicate $(Ex)(Ed)P(x, y, r, e, d)$ has degree at most $\underline{0}'$. (Recall that $\underline{0}'$ is the greatest member of the set of degrees of 1-quantifier forms.) Case 1.1 is true at stage s if and only if

$$(Ex)(Ed)P(x, y_s, r_s, e, d) \quad ,$$

where $(s)_0 = i$, $(s)_1 = e$, $r_s = \mu n(a_{i,s}(n)$ is undefined) and y_s is the weight of a_s^i. If Case 1.1 is true, then b_s is the unique, finite, partial function whose weight is

$$(\mu x)(Ed)P(x, y_s, r_s, e, d) \quad ,$$

and $a_i(r_s) = 1 + \{e\}^{b_s}(r_s)$. If Case 1.1 is false, then $a_i(r_s) = 0$. Thus we pass from a_s^i and $a_{i,s}$ to a_{s+1}^i and $a_{i,s+1}$ with the aid of a predicate of degree at most $\underline{0}'$. In other words, we can express a_{s+1}^i and $a_{i,s+1}$ in terms of a_s^i and $a_{i,s}$ with the aid of a predicate of degree at most $\underline{0}'$. In the light of the concluding remarks of Section 1, it is clear that $\lambda in|A(i, n)$ has degree at most $\underline{0}'$.

For each i and n, let

$$b_i(n) = \prod_{j<i} p_j^{a_j(n)} \quad .$$

Thus a_j is recursive in b_i when $j < i$; in addition, b_i is recursive in b_{i+1}. Since a_j is recursive in a^i when $j \neq i$, and since a_i is not recursive in a^i, it follows that a_i is not recursive in $a_0, a_1, \ldots, a_{i-1}$. But then, b_i is not recursive in b_{i-1} for all positive i.

We are now ready to construct two functions, d_0 and d_1, whose degrees have no greatest lower bound. Each b_i will be recursive in both d_0 and d_1. Any function that is recursive in both d_0 and d_1 will be recursive in one of the b_i's. It will follow from these last two facts that the degrees of d_0 and d_1 have no greatest lower bound. Since $\lambda in|A(i, n)$ has degree less than or equal to $\underline{0}'$, it will be possible to construct d_0 and d_1 in such a manner that their degrees will be less than or equal to $0''$.

The definition of d_0 and d_1 proceeds by stages. At stage $s \geq 0$, we define finitely or infinitely many values of d_0 and d_1. Before we proceed with stage s, we state an induction hypothesis concerning what has happened in the course of the construction prior to stage s:

(L1) for each $i < 2$, there are only finitely many n such that $d_i(n)$ has been defined and n is not a power of a prime p_j for some $j < s$.

(L2) for each $i < 2$, if $d_i(p_j^m)$ has been defined for infinitely many m, then $2^j < s$ and there exists a $t(j)$ such that $d_i(p_j^{t(j)+n})$ has been defined and set equal to $b_j(n)$ for all $n \geq 0$.

There are only three cases at stage s.

CASE 1. $2^j = s$. The object of this case is to insure that b_j is recursive in both d_0 and d_1 when the construction is completed. It follows from (L2) that there is a t such that $d_i(p_j^{t+n})$ has not het been defined for any $n \geq 0$ and $i < 2$; let $t(j)$ be the least such t. We define

$$d_i(p_j^{t(j)+n}) = b_j(n)$$

for all $n \geq 0$ and $i < 2$.

CASE 2. $(s)_0 = 0$ and $(s)_3 = 1$. Let $(s)_1 = e$ and $(s)_2 = f$. The purpose of this case it to insure that any function recursive in d_0 with Gödel number e and recursive in d_1 with Gödel number f will be recursive in one of the b_i's. For each $i < 2$, let d_i^s be the unique partial function whose domain consists of all n such that $d_i(n)$ has been defined prior to stage s and whose values are given by $d_i^s(n) = d_i(n)$. For each $i < 2$, the notion of weight well-orders the set of all finite extensions of d_i^s. If b is a finite extension of d_i^s, then the weight of b is an effective encoding of those n such that $b(n)$ is defined and $d_i^s(n)$ is not defined and also of the values of $b(n)$ for those n for which $b(n)$ is defined and $d_i^s(n)$ is not.

CASE 2.1. There exists an n such that for any finite extensions b of d_0^s and g of d_1^s, either $\{e\}^b(n)$ or $\{f\}^g(n)$ is undefined. We proceed directly to stage $s + 1$.

CASE 2.2. Case 2.1 does not hold, and there exist an n and finite extensions, b and c, of d_0^s such that

$$\{e\}^b(n) \quad \text{and} \quad \{e\}^c(n)$$

are defined but are not equal. Among all such triads (n, b, c) there is a unique one with the property that

$$2^n \cdot 3^{\text{weight of } b} \cdot 5^{\text{weight of } c}$$

has the least possible value; let it be denoted by (n_s, b^s, c^s). Since Case 2.1 does not hold, there exists a finite extension g of d_1^s such that $\{e\}^g(n_s)$ is defined; let g^s be the one of least weight. Let h^s be that member of $\{b^s, c^s\}$ of least weight such that

$$\{e\}^{h^s}(n_s) \neq \{f\}^{g^s}(n_s) \quad .$$

We define $d_0(n)$, $d_1(n)$ respectively, to be $h^s(n)$, $g^s(n)$ respectively, for each n for which $h^s(n)$, $g^s(n)$ respectively, is defined and $d_0^s(n)$, $d_1^s(n)$ respectively, is not defined.

CASE 2.3. Neither Case 2.1 nor Case 2.2 holds. We proceed directly to stage $s + 1$.

CASE 3. s is otherwise. Let r_s be the least n such that $d_0^s(n)$ is undefined, and let v_s be the least n such that $d_1^s(n)$ is undefined. We define $d_0^s(r_s) = d_1^s(v_s) = 0$.

That completes the construction. By Case 3, d_0 and d_1 are defined everywhere. By Case 1, each b_i is recursive in both d_0 and d_1. We show that the degrees of d_0 and d_1 have no greatest lower bound. Let w be a function whose degree is a lower bound on the degrees of d_0 and d_1. We will find a b_i whose degree is greater than the degree of w. Let e and f be Gödel numbers such that

$$w = \{e\}^{d_0} = \{f\}^{d_1} \quad .$$

Let s be such that $(s)_0 = 0$, $(s)_3 = 1$, $(s)_1 = e$ and $(s)_2 = f$. Thus Case 2 holds at stage s. If Case 2.1 holds at stage s, then there exists an n such that either

$$\{e\}^{d_0}(n) \quad \text{or} \quad \{f\}^{d_1}(n)$$

is undefined. It follows that Case 2.1 does not hold at stage s. If

Case 2.2 holds, then there exists an n (namely, n_s) such that

$$w(n) = \{e\}^{d_0}(n) \neq \{f\}^{d_1}(n) = w(n) \quad .$$

It follows that Case 2.3 holds at stage s. Thus for each n, we have:

(i) there is a finite extension b of d_0^s such that $\{e\}^b(n)$ is defined;

(ii) if b and c are finite extensions of d_0^s such that $\{e\}^b(n)$ and $\{e\}^c(n)$ are both defined, then $\{e\}^b(n) = \{e\}^c(n)$.

We will use (i) and (ii) to show that $w = \{e\}^{d_0}$ is recursive in b_s. Let D_s be the domain of d_0^s. Then

$$D_s = F \cup \left\{p_j^{n+t(j)} \middle| 2^j < s \;\&\; n \geq 0\right\} \quad ,$$

where F is some finite set. This last is clear from (L1) and (L2). Thus D_s is recursive. It also follows from (L2) that the values of d_0^s are effectively computable from b_s, since b_j is recursive in b_s for all $j < s$. For each n, let $z(n)$ be

$$\mu x(Ey)_{y<x}(Eb)\Big(b \text{ is a finite extension of } d_0^s$$
$$\& \;(m)_{m<y}(b(m) \text{ is defined}) \;\&\; x \text{ is the weight of } b$$
$$\& \;T_1^1\big(\tilde{b}(y),\, e,\, n,\, y\big)\Big) \quad .$$

It follows from (i) that $z(n)$ is well-defined. It is clear that z is computable from D_s, d_0^s. But then z is recursive in b_s. For each n, let

$$v(n) = \{e\}^b(n) \quad ,$$

where b is the unique finite extension of d_0^s of weight $z(n)$. Then v is recursive in b_s. Since $w(n) = \{e\}^{d_0}(n)$, it follows from (ii) that $w(n) = v(n)$. Thus w is recursive in b_s. Recall that b_s is recursive in b_{s+1} but b_{s+1} is not recursive in b_s. Then w is recursive in b_{s+1} and b_{s+1} is not recursive in w. Since b_{s+1} is recursive in both d_0 and d_1, it follows that the degree of w is not the greatest lower bound of the degrees of d_0 and d_1.

It remains only to see that the degrees of d_0 and d_1 are at most $\underline{0}''$. Let $\lambda in|B(i, n)$ be a function such that for each i, $\lambda n|B(i, n) = b_i$. It is easily checked that $\lambda in|B(i, n)$ is recursive in $\lambda in|A(i, n)$. We saw above that the degree of $\lambda in|A(i, n)$ was at most $\underline{0}'$.

If we can show that both d_0 and d_1 are recursive in a composition of $\lambda in|B(i, n)$ and a predicate of degree $\leq \underline{0}''$, then it will be clear that the degrees of d_0 and d_1 are at most $\underline{0}''$. Consider what happens at stage s of the simultaneous definition of d_0 and d_1. If Case 1 holds, then $t(j)$ is obtained effectively from d_0^s and d_1^s, and d_0^s and d_1^s are finitely extended with the help of $\lambda in|B(i, n)$. Suppose Case 2 holds. We can tell which of the three cases, 2.1, 2.2, or 2.3, holds by composing $\lambda in|B(i, n)$ and a predicate of degree $\underline{0}''$. (We do not exhibit the predicate since it is similar in nature to the one-quantifier form occurring in our argument that $\lambda in|A(i, n)$ has degree at most $\underline{0}'$.) With the help of this predicate we can single out the desired extensions of d_0^s and d_1^s. If Case 3 holds, we extend d_0^s and d_1^s in a trivial fashion. Thus we can explicitly define d_0^{s+1} and d_1^{s+1} in terms of d_0^s and d_1^s with the aid of a predicate of degree at most $\underline{0}''$ and the function $\lambda in|B(i, n)$. It follows with the help of the closing remarks of Section 1 that d_0 and d_1 are of degree at most $\underline{0}''$.

Each of the theorems of the present section was proved by means of the diagonal method. Each of the above constructions amounted to a definition of a function by induction; the function had to meet countably many requirements, and stage s of the induction was devoted to meeting the sth requirement. Godel's construction of an undecidable, arithmetical predicate and Kleene's construction of a non-recursive, one-quantifier form made similar use of the diagonal method. In Sections 4-7, 9 and 11 we will see that the diagonal method lacks the power needed to obtain results about degrees deeper than those of Sections 2 and 3.

§3. UNCOUNTABLE SUBORDERINGS OF DEGREES

In this section we study conditions which are sufficient and, in some cases, necessary for a partially ordered set to be imbeddable in the upper semi-lattice of degrees. Two partially ordered sets, M and M', are called order-isomorphic if there is a map $m \to m'$ of M onto M' such that $m \leq n$ if and only if $m' \leq n'$. A partially ordered set P is said to be imbeddable in a partially ordered set Q if P is order-isomorphic to some subset of Q. Let T be a partially ordered set of cardinality at most that of the continuum such that each member of T has at most aleph-one successors; we show T is imbeddable in the degrees if and only if each member of T has at most countably many predecessors. Let A and B be sets of degrees such that A is countable, B has cardinality less than that of the continuum, and no member of B is less than or equal to any finite union of members of A; we show there exists a degree $\underline{d}$ such that $\underline{d}$ is greater than every member of A and incomparable with every member of B. Finally, we show that if T is a partially ordered set of cardinality at most that of the continuum with the property that each member of T has at most finitely many predecessors, then T is imbeddable in the degrees. We assume complete familiarity with the arguments of Section 2.

> THEOREM 1. Let T be a partially ordered set, and let M and N be disjoint subsets of T such that M has cardinality less than that of the continuum, N is countable, and no member of N is less than any member of M. For each $n \in N$, the set $M_n = \{m | m \in M \ \& \ m < n\}$ is countable and any two members of M_n have an upper bound in M_n. Let A be a set of degrees order-isomorphic to M. Then there exists a set B of degrees such that $A \cup B$ is order-isomorphic to $M \cup N$ by means of an extension of the given order-isomorphism between A and M.

PROOF. Let $M = \{m_t | t \in V\}$, $A = \{\underline{a}_t | t \in V\}$ and $N = \{n_i | i = 0, 1, 2, \ldots\}$, and let the given order-isomorphism between A and M be such that for each t and u in the index set V, $\underline{a}_t \leq \underline{a}_n$ if and only if $m_t \leq m_u$. We will obtain a sequence of functions $\{b_i | i = 0, 1, 2, \ldots\}$ such that the sequence of degrees $\{\underline{b}_i | i = 0, 1, 2, \ldots\}$ will constitute the desired set B. For each $t \in V$, let a_t be a function of degree $\underline{a}_t$. The sequence $\{b_i\}$ must meet five sets of requirements:

(R1) b_j recursive in b_i if $n_j \leq n_i$;

(R2) a_t recursive in b_i if $m_t \leq n_i$;

(R3) b_k not recursive in b_i if $n_k \not\leq n_i$;

(R4) a_t not recursive in b_i if $m_t \not\leq n_i$;

(R5) b_i not recursive in a_t if $t \in V$.

The sequence $\{b_i\}$ of functions will be extracted from a sequence $\{B_i\}$ of sets of functions. For each i, B_i will have cardinality of the continuum, and the members of B_i will be indexed by sets of natural numbers. Let S be the set of all sequences of natural numbers such that $v = \{v(k) | k = 0, 1, 2, \ldots\}$ is a member of S if and only if for each $k \geq 0$, $1 \leq v(k) \leq 2^k$ and $2v(k) - 1 \leq v(k+1) \leq 2v(k)$. For each i there will be a one-to-one correspondence between B_i and S which will serve to index the members of B_i; for each $v \in S$, B_i^v will denote the unique member of B_i whose index is v. The sequence $\{B_i\}$ of sets will be defined in such a way that for each $v \in S$, the sequence $\{B_i^v\}$ of functions will meet three requirements: B_j^v recursive in B_i^v if $n_j \leq n_i$; a_t recursive in B_i^v if $m_t \leq n_i$; and B_k^v not recursive in B_i^v if $n_k \not\leq n_i$. Thus a continuum of sequences of functions will be available that satisfy (R1), (R2), and (R3). Of this continuum, a subcontinuum will be found to satisfy (R4) and (R5).

Each B_i will owe its definition to a set of integer-valued functions

$$\left\{Q_{ink} | n = 0, 1, 2, \ldots; k = 1, 2, \ldots, 2^n\right\} ;$$

each Q_{ink} will have as its domain of arguments a finite non-empty set R_{in} of natural numbers. The least member of $R_{i,n+1}$ will be the immediate successor of the greatest member of R_{in}; thus the sequence

$\{R_{in} | n = 0, 1, 2, \ldots\}$ will constitute a partition of the natural numbers into consecutive, finite sets. A function f will be a member of B_i if and only if there is a $v \in S$ such that for each $n \geq 0$ and $r \in R_{in}$,

$$f(r) = Q_{inv(n)}(r) \quad ;$$

if this is the case we will say $f = B_i^v$. During the course of the construction we will say $B_i^v(m)$ is defined if and only if for some n, $Q_{inv(n)}(m)$ is defined. Note that if $B_i^v(m)$ is defined by and and hence equal to $Q_{inv(n)}(m)$, then $B_i^w(m)$ is defined for all $w \in S$ with the property that $w(n) = v(n)$.

Since for each $n_i \in N$, M_{n_i} is countable, we write

$$M_{n_i} = \{m_{ij} | j = 0, 1, 2, \ldots\} = \{t | t < n_i \ \& \ t \in M\} \quad .$$

For each i and j, let a_{ij} be a function whose degree is that member of A corresponding to m_{ij} under the given order-isomorphism between A and M. The construction of the functions $\{Q_{ink}\}$ takes the form of a definition by induction. At stage s of the construction, either or both of two kinds of activity may take place:

(1) for some i and n the functions $\{Q_{ink} | 1 \leq k \leq 2^n\}$ and their common set of arguments R_{in} are defined;

(2) irrevocable commitments are made concerning the values to be taken by some of the functions $\{Q_{ink}\}$ when it becomes time to define them. All commitments will be capable of expression by one of two utterances:

(a) for all v, as soon as one of $B_j^v(m)$ and $B_i^v(p_j^{m+r(ij)})$ is assigned the value it must take when it is defined, the other must be assigned the same value and must take that value when it is defined;

(b) if $6 \cdot p_j^{m+t(ij)}$ is put in R_{in} when R_{in} is defined, then for all k such that $1 \leq k \leq 2^n$, $Q_{ink}(6 \cdot p_j^{m+t(ij)})$ must be assigned the value $a_{ij}(m)$ and must take that value when it is defined.

All commitments will be honored at the earliest possible stage. We will describe the status of $p_j^{m+r(ij)}$, $6 \cdot p_j^{m+t(ij)}$ respectively in utterance (a), (b) respectively, by saying that $p_j^{m+r(ij)}$, $6 \cdot p_j^{m+t(ij)}$ respectively, has been committed to a reserved, closed respectively, classification in the ith partition. We will say that $B_i^v(m)$ has received a value if it has been defined or if it has been assigned a value it must take when it is defined. If a commitment of type (a) has been made at stage s, then we will say the first one of $B_j^v(m)$ and $B_i^v(p_j^{m+r(ij)})$ to receive a value induces the value of the other. Before we proceed with stage s of the construction, we must state an induction hypothesis concerning what has happened prior to stage s:

(H1) Only a finite number of the sets $\{R_{in} | i \geq 0, n \geq 0\}$ have been defined. For each i and n, if R_{in} has been defined, then R_{ik} has been defined for all $k \leq n$ and Q_{ing} has been defined for all g such that $1 \leq g \leq 2^n$. There are only finitely many triples (i, j, m) such that for some v, p_j^m was committed to a reserved classification in the ith partition and $B_i^v(p_j^m)$ subsequently received a value prior to stage s.

(H2) There are only finitely many i such that for some m, m has been committed to a reserved or closed classification in the ith partition. All natural numbers committed to a reserved, closed respectively, classification in some partition are of the form p_j^m, $6 \cdot p_j^m$ respectively, where $m > 0$ and j is restricted to a finite range of values. For each i and j: if p_j^m, $6 \cdot p_j^m$ respectively, has been committed to a reserved, closed respectively, classification in the ith partition, then there is a positive integer r(ij), t(ij) respectively, such that every natural number of the form p_j^m $(m \geq r(ij))$, $6 \cdot p_j^m$ $(m \geq t(ij))$ respectively, has been committed similarly and such that no natural number of the form p_j^m $(m < r(ij))$, $6 \cdot p_j^m$ $(m < t(ij))$ respectively, has been committed similarly: furthermore, for each $v \in S$ and natural number m, $B_j^v(m)$ and $B_i^v(p_j^{m+r(ij)})$ are joined together by a commitment of type (a) and if $6 \cdot p_j^{m+t(ij)}$ has been put in R_{in} prior to stage s, then

$$Q_{ink}(6 \cdot p_j^{m+t(ij)}) = a_{ij}(m)$$

for all k such that $1 \leq k \leq 2^n$.

(H3) For each i and j, a natural number of the form p_j^m, $6 \cdot p_j^m$ respectively, has been committed to a reserved, closed respectively, classification in the ith partition only if $2 \cdot 3^j \cdot 5^i < s$ and $j \neq i$, $2^2 \cdot 3^j \cdot 5^i < s$ respectively.

At stage s there are five possible cases. Cases 1-3 correspond to (R1) - (R3).

CASE 1. $s = 2 \cdot 3^j \cdot 5^i$ and $n_j < n_i$. We claim that there are only finitely many m such that for some v, $B_i^v(p_j^m)$ has received a value prior to stage s. The first part of (H1) tells us that there are only finitely many m such that for some v, $B_i^v(p_j^m)$ was defined prior to stage s. Suppose $B_i^v(p_j^m)$ had its value induced prior to stage s. It follows from (H2) and (H3) that p_j^m was not committed to a reserved or closed classification in the ith partition prior to stage s. But then there must be a k such that $B_k^v(p_i^t)$ $(t = p_j^m)$ received a value prior to stage s and thereby induced the value of $B_i^v(p_j^m)$. Thus p_i^t was committed to a reserved classification in the kth partition and $B_k^v(p_i^t)$ subsequently received a value prior to stage s. The last part of (H1) says that there are only finitely many triples (k, i, t) such that for some v, p_i^t was committed to a reserved classification in the kth partition and $B_k^v(p_i^t)$ subsequently received a value prior to stage s. Consequently, there are only finitely many m such that for some v, $B_i^v(p_j^m)$ had its value induced prior to stage s. The only way $B_i^v(p_j^m)$ can receive a value is to have it induced or defined. That proves our claim. Let $r(ij)$ be the least positive r such that for all $m \geq 0$ and all v, $B_i^v(p_j^{m+r})$ has not received a value prior to stage s. It follows from (H2) and (H3) that for all m, p_j^m has not yet been committed to a reserved or closed classification in the ith partition. We now commit $p_j^{m+r(ij)}$ to a reserved classification in the ith partition for all $m \geq 0$: for all v, as soon as one of $B_j^v(m)$ and $B_i^v(p_j^{m+r(ij)})$ receives a value, the other must receive the same value; if the first has received a value prior to stage s, the other must receive that value now.

CASE 2. $s = 2^2 \cdot 3^j \cdot 5^i$. We claim that there are only finitely many m such that for some v, $B_i^v(6 \cdot p_j^m)$ has received a value prior to stage s. The only way $B_i^v(6 \cdot p_j^m)$ can receive a value is to have it induced or defined. Our claim is easily proved by means of the argument given in Case 1. Let $t(ij)$ be the least, positive t such that for $m \geq 0$ and all v, $B_i^v(6 \cdot p_j^{m+t})$ has not received a value prior to stage s. Of course $6 \cdot p_j^m$ has not yet been committed to a reserved or closed classification in the ith partition for any m. We now commit $6 \cdot p_j^{m+t(ij)}$ to a closed classification in the ith partition for all $m \geq 0$: if $6 \cdot p_j^{m+t(ij)}$ is put in R_{in} at some future stage, then at that stage, $Q_{ink}(6 \cdot p_j^{m+t(ij)})$ must be set equal to $a_{ij}(m)$ for all k such that $1 \leq k \leq 2^n$.

Before we proceed with Case 3, we fix i and v and consider what is meant by a finite extension of B_i^v at stage s. It follows from (H1) and (H2) that there are infinitely many m such that $B_i^v(m)$ has not received a value prior to stage s. Let $m_0 < m_1 < \ldots < m_{z-1}$ be the first z arguments of B_i^v which have not received a value prior to stage s. Then a finite extension of B_i^v at stage s might consist simply of assigning values to $B_i^v(m_j)$ $(j < z)$. Unfortunately, the situation is complicated by the possibility of induced values. For example, the assignment of a value to $B_i^v(m_0)$ might induce some value of B_j^v which might induce some value of B_k^v which might induce a value for $B_i^v(m_1)$. Thus it may not be possible to assign values independently to $B_i^v(m_0)$ and $B_i^v(m_1)$. To make matters still worse, the assignment of a value to $B_i^v(m_0)$ might induce the value of $B_i^v(m)$ for infinitely many m.

Suppose we assign a value to $B_i^v(m)$. This assignment may induce values in two different ways. First, the value of $B_j^v(p_i^{m+r(ji)})$ may be induced; second, m may be of the form $p_k^{n+r(ik)}$ and the value of $B_k^v(n)$ may be induced. It follows from (H2) that the sets

$$\left\{j \mid \text{value of } B_j^v(p_i^{m+r(ji)}) \text{ induced by } B_i^v(m)\right\}$$

and

$$\left\{(k, n) \mid \text{value of } B_k^v(n) \text{ induced by } B_i^v(m) \ \& \ m = p_k^{n+r(ik)}\right\}$$

are finite. However, each value thus induced may itself induce a value in two different ways. Let $a_0, a_1, \ldots, a_t$ be a finite sequence of values such that $a_0 = B_i^v(m)$ and such that for each $k < t$, a_k induces the value of a_{k+1} in one of the two ways described above. We call $a_0, a_1, \ldots, a_t$ a chain of induced values of length t. We claim that $t \leq 2s$. Suppose a_k is an induced value of the first kind; then it is easily seen that a_{k+1} cannot be an induced value of the second kind. Thus to prove our claim, we need only show that if all the induced values in a chain are of the same kind, then the chain has length at most s. But this last follows the fact that there are only finitely many k such that for some n, n has been committed to a reserved classification in the kth partition (H2). Since each chain has length at most $2s$ and since each induced value can induce only finitely more values in each of the two ways, it follows that the set

$$\left\{(k, n) \mid B_k^V(n) \text{ receives a value determined by the value assigned to } B_i^V(m) \text{ and by commitments made prior to stage } s\right\}$$

is finite. Thus the assignment of a value to $B_i^V(m)$ has only finitely many repercussions despite the commitments we are forced to honor immediately. Note that it is perfectly possible for the assignment of a value to $B_i^V(m_0)$ to determine a value for $B_i^V(m_1)$. We cannot rule out such an event.

By a finite extension of B_i^V of length z at stage s we mean not only the receiving of values by the first z argument of B_i^V which have not received values prior to stage s, but also the assignment of all values thereby determined in any of the functions $\{B_j^V \mid j = 0, 1, 2, \ldots\}$ as a result of commitments. Of course the values assigned to $B_i^V(m_j)$ $(j < z)$ must be consistent with our commitments, since it may not be possible to assign values to $B_i^V(m_j)$ $(j < z)$ independently.

CASE 3. $s = 2^3 \cdot 3^k \cdot 5^e \cdot 7^i$ and $n_k \not\leq n_i$. Let $q(2)$, $q(1)$ respectively, be the least n such that R_{kn}, R_{in} respectively, has not been defined prior to stage s. Let u_2, u_1 respectively, be the greatest member of $R_{k,q(2)-1}$, $R_{i,q(1)-1}$ respectively, if $q(2)$, $q(1)$ respectively, is greater than 0; otherwise, let u_2, u_1 respectively, be -1. At stage s we say a partial function f is a finite extension of B_i^V if $f(m) = B_i^V(m)$ for all m for which B_i^V has received a value prior to stage s and if the act of putting $B_i^V(m)$ equal to $f(m)$ for all m for which B_i^V has not received a value prior to stage s would constitute a finite extension of B_i^V as described above. We require f to be complete in the following sense: if $f(m)$ is defined and if assigning the value $f(m)$ to $B_i^V(m)$ induces the value of $B_i^V(n)$, then $f(n)$ is also defined and is equal to the induced value of $B_i^V(n)$. Thus the partial function f assigns consistent values to the first z arguments of B_i^v which have not received values prior to stage s, and in addition, assigns all values thereby induced in B_i^v. We define the weight of f as we did in the proof of Theorem 1 of Section 2. The partial function $\{e\}^f$, where f is a partial function, was defined in Section 1.

CASE 3a. $q(1) = q(2) = q$. For each g such that $1 \leq g \leq 2^q$, we will define Q_{kqg} and Q_{iqg} in such a way that for all $v \in S$ with the property that $v(q) = g$, we will have B_k^v not recursive in B_i^v with Gödel number e when the construction is complete. For each g such that $1 \leq g \leq 2^q$, let $w(g)$ be the unique member of S defined by

$$w(g)(q) = q \quad ;$$
$$(k)\Big(k \geq q \rightarrow w(g)(k+1) = 2w(g)(k)\Big).$$

Let r be the least n such that $p_k^n > u_2$ and such that $B_k^v(p_k^n)$ has not received a value prior to stage s; r is well-defined by (H3). If there is a partial function f which is a finite extension of $B_i^{w(g)}$ and which has the property that

$$\{e\}^f(p_k^r) \tag{1}$$

is defined, we say g is not null. If g is not null, let the least (by weight) such finite extension of $B_i^{w(g)}$ be the needed one at stage s, let z_g be the resulting value of (1), and let t_g be the greatest argument of $B_i^{w(g)}$ which receives a value as a result of making the needed finite extension. If g is null, let t_g be $u_1 + 1$. We define:

$$t = \max\left\{t_g \mid 1 \leq g \leq 2^q\right\} + p_k^r \quad ;$$

$$R_{iq} = \left\{n \mid u_1 < n < t\right\} \quad ; \quad R_{kq} = \left\{n \mid u_2 < n < t\right\} \quad .$$

For each g and h such that $1 \leq g \leq 2^q$ and $h \in R_{iq}$, we define:

$$Q_{iqg}(h) = \begin{cases} \text{value received at some stage prior to stage } s \\ \text{value received above at stage } s \text{ (g not null)} \\ 0 \text{ otherwise.} \end{cases} \tag{2}$$

It follows from (H3) that for any $v \in S$, $B_k^v(p_k^r)$ could not have had its value induced as a consequence of the finite extensions described by (2). Thus we are at liberty to define:

$$Q_{kqg}(h) = \begin{cases} 1 \dot{-} z_g & \text{if } g \text{ is not null and } h = p_k^r \\ \text{value received at stage } s \text{ or prior to stage } s \\ 0 & \text{otherwise} \end{cases}$$

for each h and g such that $h \in R_{kq}$ and $1 \leq g \leq 2^q$.

CASE 3b. $q(1) \neq q(2)$. Suppose $q(1) > q(2)$. We first define $R_{k,q(2)+h} = \{u_2 + h\}$ and $Q_{k,q(2)+h,g}(u_2 + h)$ = value received at some stage prior to $\overline{s, 0}$ otherwise whenever $q(2) < q(2) + h \leq q(1)$ and $1 \leq g \leq 2^{q(2)+h}$, and then we proceed as in Case 3a.

CASE 4. $(s)_0 = 4$, $(s)_1 = k$ $(s)_2 = e_1$ and $(s)_3 = e_2$. Let q be the least n such that R_{kn} has not been defined prior to stage s. If $q > 0$, let u be the greatest member of $R_{k,q-1}$; otherwise, let u be -1. Define $w(g)$ as in Case 3a. Let $D = \{2^a \cdot 3^b | 1 \leq a < b \leq 2^q\}$. For each $d \in D$, let $d(0) = (d)_0$ and $d(1) = (d)_1$. For each $d \in D$ we will finitely extend $B_k^{w(d(0))}$ and $B_k^{w(d(1))}$ in such a way that when the construction is complete, either

$$(3) \quad U\Big(\mu y T_1^1\Big(\widetilde{B}_k^{w(d(0))}(y),\ e_1,\ m,\ y\Big)\Big) = U\Big(\mu y T_1^1\Big(\widetilde{B}_k^{w(d(1))}(y),\ e_2,\ m,\ y\Big)\Big)$$

will be false for some m or at least one side of (3) will be undefined for some m or the left side of (3) will be a function recursive in a_{k0}, a_{k1}, ..., a_{ks}. The order in which these pairs of finite extensions will occur is the natural order of D. Thus we begin by extending $B_k^{w(1)}$ and $B_k^{w(2)}$, since $2^1 \cdot 3^2$ is the least member of D. Then we make a further extension of $B_k^{w(1)}$ and extend $B_k^{w(3)}$, and so on. Thus for each g, $B_k^{w(g)}$ will suffer $2^q - 1$ successive, finite extensions, and the union of these will be the needed finite extension of $B_k^{w(g)}$ at stage s. Let $d \in D$ and suppose that for each $n \in \{i | i < d\} \cap D$, we have extended $B_k^{w(n(0))}$ and $B_k^{w(n(1))}$ according to the above plan. We now dispose of d. Thus we extend $B_k^{w(d(0))}$ and $B_k^{w(d(1))}$; we may have already extended $B_k^{w(d(0))}$ for the sake of some $n < d$; in that event, we will further extend it. We regard $B_k^{w(d(1))}$ similarly.

CASE 4a. There is an m such that the left side of

$$(4) \qquad U\Big(\mu y T_1^1\Big(\widetilde{f}_1(y),\ e_1,\ m,\ y\Big)\Big) = U\Big(\mu y T_1^1\Big(\widetilde{f}_2(y),\ e_2,\ m,\ y\Big)\Big)$$

is not defined for any partial function f_1 which is a finite extension of $B_k^{w(d(0))}$, or there is an m such that the right side of (4) is not defined for any partial function f_2 which is a finite extension of

$B_k^{w(d(1))}$. The needed finite extensions are both trivial (empty).

CASE 4b. There is an m, an f_1 which is a finite extension of $B_k^{w(d(0))}$ and an f_2 which is a finite extension of $B_k^{w(d(1))}$ such that both sides of (4) are defined but are not equal. To each such triple (m, f_1, f_2) we assign the weight $2^m \cdot 3^\mu \cdot 5^v$, where μ is the weight of f_1 and v is the weight of f_2. Then there is a unique triple of least weight from which we extract the needed finite extensions of $B_k^{w(d(0))}$ and $B_k^{w(d(1))}$. It follows that when the construction is complete, there will be an m (namely, the m associated with the triple of least weight) such that both sides of (3) are defined but are not equal.

CASE 4c. (4a) and (4b) are false. Then for each m the left side of (4) is defined for some f_1 which is a finite extension of $B_k^{w(d(0))}$, and for each m all finite extensions of $B_k^{w(d(0))}$ which define the left side of (4) give it the same value. This means that when the construction is complete, the left side of (3) will have the property that if it is a function defined for all m, then its values are completely determined by that portion of $B_k^{w(d(0))}$ which has been determined prior to stage s. Fix m and suppose the left side of (3) is defined when the construction is complete. To determine the value of the left side of (3), we merely take any finite extension of $B_k^{w(d(0))}$ at stage s which defines the left side of (4) and insert that finite extension in the left side of (4); the result will be the value of the left side of (3). The hypothesis of Case (4c) tells us that the finite extension we need does exist and that it does not matter what finite extension we take. Suppose that when the construction is complete, the left side of (3) is a function defined for all m; let g be that function. Then g is recursive in $B_k^{w(d(0))}$ with Gödel number e_1. But as we have just seen g is actually recursive in that portion of $B_k^{w(d(0))}$ determined prior to stage s. We now find ourselves in the same situation we encountered in Case 2.3 of the second half of the proof of Theorem 3 of Section 2. Let B_s be the unique partial function whose domain consists of all m such that $B_k^{w(d(0))}(m)$ has received a value prior to stage s and whose values are given by $B_s(m) = B_k^{w(d(0))}(m)$. Then g is computable from B_s, D_s, where D_s is the domain of B_s. It follows from (H1) and (H2) that B_s and D_s are each recursive in

$a_{k0}, a_{k1}, \ldots, a_{ks}$. (In fact, D_s is recursive.) But then g is recursive in $a_{ko}, a_{k1}, \ldots, a_{ks}$. This last follows by an argument not essentially different from that given in the analysis of Case 2.3 of the second half of Theorem 3 of Section 2. We note only that (H1)-(H3) provide us with a picture of the stage of $B_k^{w(d(0))}$ at stage s, a picture which consists of the functions $a_{ko}, a_{k1}, \ldots, a_{ks}$, and finitely many remarks about commitments.

Thus we dispose of $d \in D$ as described in (4a), (4b), and (4c). In Case (4a) and (4c), we make trivial (null) extensions of $B_k^{w(d(0))}$ and $B_k^{w(d(1))}$; Case (4b) is where the non-trivial extensions take place. After all the members of D have been exhausted in this manner, we define z to be the greatest integer which has received a value during the above series of finite extensions; that is, z is the greatest argument among all those arguments of the functions $\left\{B_i^v | v = w(1), w(2), \ldots, w(2^q); \ i \geq 0\right\}$ which received values during the above series of finite extensions. Remember that we include induced values. If all the extensions were trivial we take z to be $u + 1$; u was defined at the beginning of Case 4. We define

$$R_{kq} = \{h | u < h \leq z\}, \text{ and}$$

$$Q_{kqg}(h) = \begin{cases} \text{value received above at stage } s \text{ or prior to } s \\ 0 \quad \text{otherwise} \end{cases}$$

for all h and g such that $h \in R_{kq}$ and $1 \leq g \leq 2^q$. In our definition of D we assumed that $q > 0$. If $q = 0$, we simply ignore (4a), (4b), and (4c), and proceed directly to stage $s + 1$.

CASE 5. s is otherwise. For each i, let $m(i)$ be the least n such that R_{in} has not been defined prior to stage s, and let u_i be the greatest member of $R_{i,m(i)-1}$ if $m(i) > 0$ and -1 otherwise. If $1 \leq g \leq 2^{m(i)}$, then we write

$$g = c_0^g \cdot 2^0 + c_1^g \cdot 2^1 + \ldots + c_{m(i)}^g \cdot 2^{m(i)} \quad .$$

For each $i \leq s$, we define

$$R_{im(i)} = \left\{h | u_i < h \leq p_i^{m(i)+v(i)}\right\} \quad ,$$

where $v(i)$ is the least n such that there is a $v \in S$ with the property that $B_i^v(p_i^n)$ has not received a value prior to stage s. For each i, g, and h such that $i \leq s$, $1 \leq g \leq 2^{m(i)}$ and $h \in R_{im(i)}$, we define

$$(5)\qquad (Q)_{im(i)g}(h) = \begin{cases} c^g_m \text{ if } h = p_i^{v(i)+m} \text{ and } 0 \leq m \leq m_i \\ \text{value received at or prior to stage } s \\ 0 \text{ otherwise.} \end{cases}$$

Note that the finite extensions made in (5) may induce certain values, but that there can be no conflict between induced values and values defined directly in (5) because $p_i^{v(i)+m}$ cannot have been committed to a reserved classification in the ith partition by virtue of (H3). Thus the assignments of values made in (5) are consistent with our commitments.

The induction hypothesis for stage $s+1$ is obtained by replacing s by $s+1$ in (H1), (H2), and (H3). In the light of the remarks on finite extensions made between Cases 2 and 3, it is easy to see that the induction hypothesis for stage $s+1$ holds.

The precautions taken in Case 5 insure that for each pair (u, v) of distinct members of S and for each k, the functions B^u_k and B^v_k are distinct. Thus for each k, the set $\{B^v_k | v \in S\}$ has cardinality equal to that of the continuum. For each $v \in S$, we have a collection $\{B^v_i | i \geq 0\}$ of functions which satisfy (R1), (R2), and (R3): B^v_j is recursive in B^v_i if $n_j \leq n_i$ by Case 1; a_t is recursive in B^v_k if $m_t \leq n_k$ by Case 2; and B^v_k is not recursive in B^v_i if $n_k \not\leq n_i$ by Case 3.

For each k, we define

$$R_k = \left\{v | v \in S \;\&\; (Et)(t \in V \;\&\; a_t \text{ recursive in } B^v_k \;\&\; m_t \not\leq n_k)\right\} ,$$

and show R_k has cardinality less than that of the continuum. Since V has cardinality less than that of the continuum, it is sufficient to show that for each $t \in V$, there is at most one $v \in S$ with the property that a_t is recursive in B^v_k and $m_t \not\leq n_k$. Suppose this last is false. Thus we have t, u and v such that $u \neq v$, a_t is recursive in B^u_k with Gödel number e_1, a_t is recursive in B^v_k with Gödel number e_2, and $m_t \not\leq n_k$. Let z be the least n such that $u(k) \neq v(k)$ for all $k \geq n$. Let s be such that $(s)_1 = k$, $(s)_2 = e_1$, $(s)_3 = e_2$ and $q > z$, where q is the least n such that R_{kn} has not been defined prior to stage s. Then Case 4 holds at stage s. By the definition of z, there must be a

$d \in D$ such that

$$\{d(0), d(1)\} = \{u(q), v(q)\} \quad .$$

The measures we took in Case 4 guarantee that either

$$(6) \qquad U\Big(\mu y T_1^1\Big(\tilde{B}_k^u(y), e_1, m, y\Big)\Big) = U\Big(\mu y T_1^1\Big(\tilde{B}_k^v(y), e_2, m, y\Big)\Big)$$

will be false for some m or at least one side of (6) will be undefined for some m or the left side of (6) will be recursive in $a_{k0}, a_{k1}, \ldots, a_{ks}$. But both sides of (6) are equal to a_t; consequently, a_t is recursive in $a_{k0}, a_{k1}, \ldots, a_{ks}$. It follows from the hypothesis of Theorem 1 that each finite subset of M_k has an upper bound in M_k. Let m_g be a member of M_k with the property that for all $i \leq s$, $m_{ki} \leq m_g$. Since $m_t \not\leq n_k$ and $m_g < n_k$, we must have $m_t \not\leq m_g$. But this is absurd, because if $\underline{a}_g$ is the element of A corresponding to m_g under the given order-isomorphism between A and M, then a_{ki} is recursive in a_g for all $i \leq s$, a_t is recursive in a_g, and $m_t \leq m_g$.

For each k, let $S_k = \Big\{v \mid v \in S \;\&\; (Et)\Big(t \in V \;\&\; B_k^v \text{ recursive in } a_t\Big)\Big\}$. Since V has cardinality less than that of the continuum, so must S_k. Let $H = \cup\{R_k \cup S_k \mid k \geq 0\}$. Since H has cardinality less than that of the continuum, it follows $S - H$ is not empty. For each $v \in S - H$, we have a sequence $\{B_i^v \mid i \geq 0\}$ of functions which satisfy (R1)-(R5). It is easy to define a unique member w of $S - H$ by induction.

That finishes the proof of Theorem 1. The main difficulty in our argument arose from the fact that we were given M as having cardinality less than that of the continuum. If M were countable, we could have proceeded quite simply in the vein of Section 2. Since we allowed the possibility of M being uncountable, we had to make use of a counting argument based on Case 4 in order to satisfy (R4) and (R5). If M, and hence V, were countable, we could have satisfied (R4) and (R5) by a direct construction similar to those occurring in Section 2.

We say a partially ordered set P is completely normal if there exists an ordinal α and a collection $\{B_\gamma \mid \gamma < \alpha\}$ of subsets of P such that $P = \cup\{B_\gamma \mid \gamma < \alpha\}$ and such that for each ordinal $\gamma < \alpha$:

(1) $\cup \{B_\delta | \delta < \gamma\}$ has cardinality less than that of the continuum;

(2) B_γ is at most countable and is disjoint from $\cup \{B_\delta | \delta < \gamma\}$;

(3) no member of B_γ is less than any member of $\cup \{B_\delta | \delta < \gamma\}$;

(4) for each $n \in B_\gamma$, the set $L_n^\gamma = \{a | a < n \ \& \ a \in \cup \{B_\delta | \delta < \gamma\}\}$

is at most countable and any two members of L_n^γ have an upper bound in L_n^γ.

We say a partially ordered set Q is normal if there exists a completely normal, partially ordered set P in which Q can be imbedded. Note that if P is normal, then each member of P has at most a countable number of predecessors, and P has cardinality at most that of the continuum.

LEMMA 2. Let P be a partially ordered set of cardinality at most aleph-one. Then P is normal if and only if each member of P has at most a countable number of predecessors.

PROOF. Let P be a partially ordered set of cardinality at most aleph-one such that each member of P has at most a countable number of predecessors. For each $p \in P$, let

$$p' = \{u | u \in P \ \& \ u \leq p\} \quad ;$$

let $P^* = \{p' | p \in P\}$ and let P' be the set of all finite unions of members of P^*. P' is an upper semi-lattice partially ordered by set-inclusion and with set-theoretic union as its join operation; furthermore, P is imbeddable in P', each member of P' has at most a countable number of predecessors, and P' has cardinality at most aleph-one. We claim P' is completely normal. Let α be the least ordinal such that the set of all ordinals less than α has the same cardinality as P', and let f be a one-one map of the ordinals less than α onto P'. For each ordinal $\gamma < \alpha$, let

$$C_\gamma = \{f(\delta) | \delta \leq \gamma\} \quad ,$$

$$C'_\gamma = \text{set of all finite unions of members of } C_\gamma \quad ,$$

$$A_\gamma = \{a | a \in P' \ \& \ (Eb)(b \in C'_\gamma \ \& \ a \leq b)\} \quad ,$$

$$B_\gamma = A_\gamma - \cup \{A_\delta | \delta < \gamma\} \quad .$$

It is readily seen that $\{B_\gamma | \gamma < \alpha\}$ is a collection of subsets of P' that possesses the properties needed to show P' is completely normal.

THEOREM 3. Let A and B be sets of degrees such that A is countable and B has cardinality less than that of the continuum; let D be the set of all degrees greater than every member of A and incomparable with every member of B. Let T be a non-empty, normal, partially ordered set. Then T is imbeddable in D if and only if no member of B is less than or equal to any finite union of members of A.

PROOF. If T is non-empty and imbeddable in D, then the necessity of the condition on A and B is clear. We show sufficiency for T normal. Let T be imbeddable in T', where T' is a completely normal, partially ordered set. Let α be an ordinal, and let $\{B_\gamma | \gamma < \alpha\}$ be a collection of subsets of T' that has the properties necessary for T' to be normal. Without loss of generality we assume that the least upper bound of any two members of A is a member of A and that $(A \cup B) \cap T'$ is empty. Let $T'' = A \cup B \cup T'$. We define a partial ordering for T'' by retaining the partial orderings of $A \cup B$ and of T' and by specifying that each member of T' is greater than every member of A and incomparable with every member of B. By means of a transfinite induction on the ordinals less than α, we construct a one-one function g from T'' into the degrees which imbed T' in D. Suppose γ is an ordinal less than α and g is already defined on

$$A \cup B \cup (\cup \{B_\delta | \delta < \gamma\})$$

in such a way that g imbeds $\cup \{B_\delta | \delta < \gamma\}$ in D and g restricted to $A \cup B$ is the identity function. In order to extend g to

$$A \cup B \cup (\cup \{B_\delta | \delta \leq \gamma\})$$

in such a way that g imbeds $\cup \{B_\delta | \delta \leq \gamma\}$ in D, we need only verify:

(1) $A \cup B \cup (\cup \{B_\delta | \delta < \gamma\})$ has cardinality less than that of the continuum;

(2) B_γ is at most countable and is disjoint from $A \cup B \cup (\cup \{B_\delta | \delta < \gamma\})$;

(3) no member of B_γ is less than any member of $A \cup B \cup (\cup \{B_\delta | \delta < \gamma\})$;

(4) for each $n \in B_\gamma$, the set $L_n^\gamma =$
$\{a \mid a < n \ \& \ a \in A \cup B \cup (\cup \{B_\delta \mid \delta < \gamma\})\}$ is countable
and any two members of L_n^γ have an upper bound in L_n^γ.

A direct application of Theorem 1 of the present section completes the proof.

COROLLARY 1. Let T be a partially ordered set of cardinality at most aleph-one. Then T is imbeddable in the upper semi-lattice of degrees if and only if each member of T has at most a countable number of predecessors.

COROLLARY 2. Let T be a partially ordered set of cardinality at most that of the continuum with the property that each member of T has at most aleph-one successors. Then T is imbeddable in the upper semi-lattice of degrees if and only if each member of T has at most a countable number of predecessors.

PROOF. The first corollary follows from the second. Let T be a partially ordered set of cardinality at most that of the continuum with the property that each member of T has at most aleph-one successors and at most countably many predecessors. We show T is normal. If the continuum hypothesis holds, then T is normal by Lemma 2. Suppose the continuum hypothesis is false. If T_1 and T_2 are subsets of T such that every member of T_1 is incomparable with every member of T_2, then we say T_1 and T_2 are incomparable subsets of T. We claim that T is the union of a collection of disjoint, mutually incomparable subsets of T such that each member of the collection has cardinality at most aleph-one. We prove this last by transfinite induction. Let α be an ordinal, and let $\{T_\gamma \mid \gamma < \alpha\}$ be a collection of disjoint, mutually incomparable subsets of T such that each T_γ has cardinality at most aleph-one; furthermore, suppose every member of T which is comparable with some member of some T_γ is a member of that T_γ; finally, suppose $T - \cup \{T_\gamma \mid \gamma < \alpha\}$ is non-empty. Let $t \in T - \cup \{T_\gamma \mid \gamma < \alpha\}$. Let $H(0) = \{t\}$; for each $n \geq 0$, let $H(n+1)$ be the set of all members of T which are comparable with some member of $H(n)$. Let

$$T_\alpha = \cup \{H(n) \mid n \geq 0\} \quad .$$

Since each member of T has at most aleph-one successors and at most countably many predecessors, it follows T_α has cardinality at most aleph-one. Since each member of T which is comparable with some member of some $T_\gamma (\gamma < \alpha)$ is a member of that T_γ, it follows that T_α is disjoint from and incomparable with $\cup \{T_\gamma | \gamma < \alpha\}$. It is clear from the definition of H that each member of T which is comparable with some member of T_α is a member of T_α.

Thus there exists an ordinal δ and a collection $\{T_\gamma | \gamma < \delta\}$ of disjoint, mutually incomparable subsets of T whose union is T with the property that each T_γ has cardinality at most aleph-one. By repeating the argument of Lemma 2 of the present section, we can obtain for each $\gamma < \delta$, a completely normal partially ordered set T'_γ such that T_γ is imbeddable in T'_γ and such that T'_γ has cardinality at most aleph-one. We can assume that the members of $\{T'_\gamma | \gamma < \alpha\}$ are disjoint. Let $T' = \cup \{T'_\gamma | \gamma < \delta\}$. We define a partial ordering for T' by retaining the partial ordering of each T'_γ and by specifying that for each pair (γ, α) of distinct ordinals less than δ, every member of T'_γ is incomparable with every member of T'_α. Then T is imbeddable in T', and T' is completely normal.

> COROLLARY 3. Let A and B be sets of degrees such that A is countable, B has cardinality less than that of the continuum, and no member of B is less than or equal to any finite union of members of A. Then there exists a degree $\underline{d}$ such that $\underline{d}$ is greater than every member of A and incomparable with every member of B.

The notion of independent degrees was defined in Section 1. We call the functions in a set independent if the degrees of the members of the set are independent. We say two functions are incomparable if their degrees are. Since Theorem 4 is a consequence solely of the methods used in the proof of Theorem 1, we are content to sketch its proof.

> THEOREM 4. Let A and B be sets of functions such that A is countable and B has cardinality less than that of the continuum. Then (1) and (2) are equivalent:
> (1) there is a set C of independent functions such

that C has cardinality of the continuum, each member of C is incomparable with every member of B, and each member of A is recursive in every member of C;
(2) no member of B is recursive in the members of any finite subset of A.

PROOF. We show (2) implies (1). Let $A = \{a_0, a_1, a_2, \ldots,\}$, and let S be defined as in Theorem 1. A set of integer-valued functions $\{Q_{nk} | n \geq 0 \ \& \ 1 \leq k \leq 2^n\}$ will be defined by stages. Each Q_{nk} will have a non-empty, finite set R_n as its domain of arguments. For each n, the least member of R_{n+1} will be the immediate successor of the greatest member of R_n. We will define a set D of functions such that f will be a member of D if and only if there is a member v of S such that for all $n \geq 0$, the restriction of f to R_n is $Q_{nv(n)}$. After straightforward modifications, the remarks made in the proof of Theorem 1 above concerning commitments and finite extensions apply here. The induction hypothesis at stage s of the construction is:

(H1) Only a finite number of the sets $\{R_n | n = 0, 1, 2, \ldots\}$ have been defined prior to stage s. If R_k has been defined prior to stage s, then for each $n \leq k$, R_n and Q_{ng} have also been defined prior to stage s for each g such that $1 \leq g \leq 2^n$.

(H2) All natural numbers committed to a closed classification prior to stage s are of the form p_j^m, where $m > 0$ and j is restricted to a finite range of values. For each j, if some natural number of the form p_j^m has been committed to a closed classification prior to stage s, then there is a positive integer t(j) such that every natural number of the form $p_j^{m+t(j)}$ $(m \geq 0)$ has been similarly committed and such that no natural number of the form p_j^m $(m < t(j))$ has been similarly committed; furthermore, for each m, if $p_j^{m+t(j)}$ has been put in R_n prior to stage s, then $Q_{ng}(p_j^{m+t(j)})$ has been set equal to $a_j(m)$ for all g such that $1 \leq g \leq 2^n$.

(H3) For all m and j, p_j^m has been committed to a closed classification only if $2 \cdot 3^j < s$.

CASE 1. $s = 2 \cdot 3^j$. The object of this case is to insure that each member of A is recursive in every member of D. Similar to Case 2 of Theorem 1.

CASE 2. $(s)_0 = 3$, $(s)_1 = e_1$ and $(s)_2 = e_2$. The purpose of this case is to guarantee that for each pair (u, v) of distinct members of S, if a function f is recursive in D^u with Gödel number e_1 and in D^v with Gödel number e_2, then f is recursive in the members of some finite subset of A. Similar to Case 4 of Theorem 1.

CASE 3. $(s)_0 = 4$ and $(s)_1 = e$. This case is needed to make the members of D independent. Let q be the least n such that R_n has not been defined prior to stage s. Let $p = 2^q$. For each g such that $1 \leq g \leq 2^q = p$, define w(g) as in Case 4 of Theorem 1. For each g we finitely extend $D^{w(1)}, D^{w(2)}, \ldots, D^{w(p)}$ in such a way that

$$D^{w(g)}(m) = U\Big(\mu y T_1^{1,1,\ldots,1}(\tilde{D}^{w(1)}(y), \ldots, \tilde{D}^{w(g-1)}(y), \tilde{D}^{w(g+1)}(y), \ldots, \tilde{D}^{w(p)}(y), e, m, y)\Big) \tag{1}$$

will be false for some m or the right side of (1) will be undefined for some m when the construction is complete. In the process each $D^{w(g)}$ is finitely extended 2^q times and the union of these consecutive 2^q extensions is the needed finite extension at stage s. This case is modeled after Cases 3 and 4 of Theorem 1.

CASE 4. s is otherwise. Define q as in Case 3. Let u be the greatest member of R_{q-1} if $q > 0$ and -1 otherwise. We define:

$$R_q = \{u + 1\}$$

$$Q_{qg}(u+1) = \begin{cases} \text{value assigned prior to stage } s \\ 0 \quad \text{otherwise,} \end{cases}$$

for all g such that $1 \leq g \leq 2^q$.

By Case 4, the members of D are well-defined. By Case 3, the members of D are independent and D has cardinality equal to that of the continuum. Let F be the set of all members of D which are comparable with some member of B. Since B has cardinality less than that of the continuum and A is countable, it follows from Case 2 that F has cardinality less than that of the continuum. D - F is the desired set C.

COROLLARY 1. There exists a set of independent degrees whose cardinality is that of the continuum.

COROLLARY 2. Let T be a partially ordered set of cardinality less than or equal to that of the continuum. If each member of T has only finitely many predecessors, then T is imbeddable in the upper semi-lattice of degrees.

PROOF. In Theorem 4 take A and B to be empty and thus obtain G, a set of independent degrees whose cardinality is that of the continuum. Let T be a partially ordered set as described in the hypothesis of Corollary 2. Let f be a one-one map of T into G. For each $t \in T$, let

$$t' = \{w \mid w \in T \ \& \ w \leq t\} \quad .$$

Since for each t, t' is finite, we write $t' = \{u_0, u_1, \ldots, u_{n(t)}\}$. For each $t \in T$, let $t'' = f(u_0) \cup f(u_1) \cup \ldots \cup f(u_{n(t)})$. Let $T'' = \{t'' \mid t \in T\}$. Then T is order-isomorphic to the set of degrees T'' by means of the map $t \to t''$.

Note that these last two corollaries are proved without any use of the axiom of choice. In the proof of Theorem 3, the axiom of choice is needed only to show F has cardinality less than that of the continuum. In Corollaries 1 and 2, F is empty because B is empty.

Theorem 3 tells us that normality is a sufficient condition for a partially ordered set to be imbeddable in the degrees. In Corollary 2 to Theorem 3, we saw that it is possible to show certain partially ordered sets of cardinality of the continuum are normal without any use of the continuum hypothesis. Unfortunately, the continuum hypothesis is closely tied to the question of which partially ordered sets are normal.

LEMMA 5. Let T be a partially ordered set of cardinality of the continuum such that any two members of T have an upper bound in T and such that any member of T has at most countably many predecessors. Then T is normal if and only if the continuum hypothesis holds.

PROOF. If the continuum hypothesis holds, then by Lemma 2 above, T is normal. Suppose T is normal. Then there exists a completely normal, partially ordered set P and a one-one map f of T into P which imbeds T in P. Let α be an ordinal and $\{B_\gamma | \gamma < \alpha\}$ be a collection of subsets of P with the properties listed in the definition of complete normality. We assume that each B_γ is non-empty. Suppose for the sake of a reductio ad absurdum that the continuum hypothesis is false. Then the set of all ordinals less than α has cardinality greater than aleph-one, since each B_γ is countable, and since T has cardinality equal to that of the continuum and is imbeddable in

$$P = \cup \{B_\gamma | \gamma < \alpha\} \quad .$$

There must exist an ordinal $\gamma^* < \alpha$ with the property that $f(T) \cap B_{\gamma^*} \neq 0$ and $f(T) \cap \cup \{B_\gamma | \gamma < \gamma^*\}$ has cardinality greater than or equal to aleph-one. Let $t \in f(T) \cap B_{\gamma^*}$. Let $d \in f(T) \cap \cup \{B_\gamma | \gamma < \gamma^*\}$. If we can show that the pair (t, d) has an upper bound in B_{γ^*}, then it will follow B_{γ^*} is uncountable, since

$$f(T) \cap \cup \{B_\gamma | \gamma < \gamma^*\}$$

is uncountable, and since each member of P has at most countably many predecessors. But then we will have attained the desired absurdity, since B_{γ^*} is at most countable according to the definition of complete normality.

We know t and d have an upper bound somewhere in P, because they are images under f of elements of T, and because any two elements of T have an upper bound in T. Let δ be the least γ such that B_γ contains an upper bound for t and d. Let $n \in B_\delta$ be such an upper bound. We show $\delta = \gamma^*$. If $\delta < \gamma^*$, then we have $t \leq n$, where $t \in B_{\gamma^*}$, $n \in B_\delta$ and $\delta < \gamma^*$. This last is ruled out by clauses (2) and (3) of the definition of complete normality. Suppose $\delta > \gamma^*$. Then clause (4) of the definition of complete normality tells us that t and d have an upper bound in B_γ for some $\gamma < \delta$. This last contradicts the definition of δ.

With the help of Lemma 5, it is immediate that normality is a necessary condition for a partially ordered set to be imbeddable in the upper semi-lattice of degrees only if the continuum hypothesis holds. After all, the upper semi-lattice of degrees itself is normal only if the continuum hypothesis holds. Nonetheless, the notion of normality is useful, since it made possible the proof of Corollary 2 to Theorem 3 without the continuum hypothesis. Corollary 2 of Theorem 4 provides us with very simple partially ordered sets which are imbeddable in the degrees but which are not provably normal without the continuum hypothesis. One such partially ordered set is the set of all finite subsets of real numbers ordered by set-inclusion.

§4. THE PRIORITY METHOD OF FRIEDBERG AND MUCHNIK

In [17] Post raised but did not answer the following problem: do all non-recursive, recursively enumerable sets have the same degree of recursive unsolvability? The solution to Post's problem was found almost simultaneously by Friedberg [1] and Muchnik [13]; these authors independently discovered a new technique which we will call the priority method. We will use the term "priority method" somewhat ambiguously to designate any method of proof which owes a large portion of its inspiration to [1] and [13]. Thus we say that Theorem 4 of [23] and Theorem 2 of [18] were proved with the help of the priority method.

Our purpose in formulating Theorem 1 of this section is to separate (insofar as is possible) the combinatorial aspects of the priority method as manifested in [1] from the recursion-theoretic aspects. We do not claim that Theorem 1 stands as a fundamental principle from which all results so far obtained by the priority method readily follow, but we do believe that Theorem 1 and its proof will be useful to anyone who wishes to develop an intuitive understanding of the workings of the priority method in all of its manifestations. We will put Theorem 1 to practical use in this section by deriving from it as corollaries the solution to Post's problem and the fact that every countable, partially ordered set can be imbedded in the upper semi-lattice of degrees of recursively enumerable sets.

A requirement $R = \{(F_i, H_i) | i \in I\}$ is a collection (possibly empty) of ordered pairs of disjoint, finite sets of natural numbers. A set T of natural numbers meets requirement R if there is an $i \in I$ such that $F_i \subseteq T$ and $H_i \cap T = 0$.

If $L = \{h_0, h_1, \ldots, h_m\}$ is a finite set of natural numbers, we define $j(L) = 2^{h_0} + 2^{h_1} + \ldots + 2^{h_m}$. The function j is a one-to-one map of the set of all finite sets of natural numbers onto the set of all natural numbers, if it is understood that $j(0) = 0$. Both j and its inverse j^{-1} are effective.

Let t be a function defined on the natural numbers with natural numbers as values. We say t enumerates requirements if for each $s \geq 0$, $j^{-1}\big((t(s))_0\big)$ and $j^{-1}\big((t(s))_1\big)$ are disjoint, finite sets. The requirements enumerated by t are denoted by $R_0, R_1, R_2, \ldots$; for each k,

$$R_k = \Big\{\Big(j^{-1}\big((t(s))_0\big),\ j^{-1}\big((t(s))_1\big)\Big)\Big|\big(t(s)\big)_2 = k\Big\} .$$

If t enumerates requirements we denote $j^{-1}\big((t(s))_0\big)$, $j^{-1}\big((t(s))_1\big)$ and $(t(s))_2$ by F^s, H^s and $g(s)$ respectively. Thus for each k,

$$R_k = \{(F^s, H^s) | g(s) = k\} .$$

A set A is said to be recursively enumerable in a function f if A is the range of a function recursive in f. With each function t which enumerates requirements, we associate a set T called the priority set of t. T is defined by stages, and as we shall see, T is recursively enumerable in t and meets every member of a certain subclass of the class of requirements enumerated by t.

Stage $s = 0$. $T_0 = 0$

Stage $s > 0$. $T_s = T_{s-1}$ if (a) or (b) or (c) is true:

(a) there is an $r < s$ such that $g(r) < g(s)$, $r > 0$, $F^r \not\subseteq T_{r-1}$, $F^r \subseteq T_r$, $H^r \cap T_{s-1} = 0$ and $H^r \cap F^s \neq 0$;

(b) there is an $r < s$ such that $g(r) = g(s)$, $r > 0$, $F^r \not\subseteq T_{r-1}$, $F^r \subseteq T_r$ and $H^r \cap T_{s-1} = 0$;

(c) $H^s \cap T_{s-1} \neq 0$. $T_s = T_{s-1} \cup F^s$ otherwise. $T = \bigcup_{s=0}^{\infty} T_s$.

For each k, we say R_k is met at stage s if $s > 0$, $k = g(s)$, $F^s \not\subseteq T_{s-1}$ and $F^s \subseteq T_s$. If R_k is met at stage s, then $H^s \cap T_s = 0$, since clause (c) must be false at stage s, and since $H^s \cap F^s = 0$ for all s.

For each k, we say R_k is injured at stage s if there is an $r < s$ such that R_k was met at stage r, $H^r \cap T_{s-1} = 0$ and $H^r \cap T_s \neq 0$.

If R_k is met at stage r and is not injured at any stage after stage r, then $F^r \subseteq T_r$ and $H^r \cap T_s = 0$ for all $s \geq r$. But then $F^r \subseteq T$ and $H^r \cap T = 0$. Thus if R_k is met at stage r and is not injured at any succeeding stage, then T meets R_k.

For each k, we say R_k is t-dense if for each finite set L, there is an $s > 0$ such that $g(s) = k$, $F^s \not\subseteq T_{s-1}$, $H^s \cap T_{s-1} = 0$ and $L \cap F^s = 0$.

The definition of T may be described as follows. We begin by setting $T_0 = 0$. At stage $s > 0$, the function t presents us with a pair (F^s, H^s) of disjoint, finite sets. This pair is of interest to us because it may represent a chance to make T meet $R_{g(s)}$. If we set $T_s = T_{s-1} \cup F^s$, and if we can manage to have $T_u \cap H^s = 0$ for all $u \geq s$, then T will meet $R_{g(s)}$. If clause (c) is true at stage s, then $H^s \cap T_{s-1} \neq 0$, and stage s clearly does not represent a chance to meet $R_{g(s)}$. If clause (b) is true, then there was a stage r prior to stage s such that $R_{g(s)}$ was met at stage r and such that $R_{g(s)}$ was not injured at any stage after r and before s; consequently we set $T_s = T_{s-1}$ with the hope of not injuring $R_{g(s)}$ at any future stage.

If clause (a) is true at stage s, then there was a stage r prior to stage s at which $R_{g(r)}$ was met, and $g(r) < g(s)$; in addition, $R_{g(r)}$ was not injured at any stage after r and before s, but if T_s is set equal to $T_{s-1} \cup F^s$, then $R_{g(r)}$ will be injured at stage s. Again we set $T_s = T_{s-1}$, because we do not wish to injure $R_{g(r)}$ at stage s for the sake of meeting $R_{g(s)}$ at stage s when $g(r) < g(s)$. Thus we have assigned a higher priority to $R_{g(r)}$ than to $R_{g(s)}$ when $g(r) < g(s)$.

LEMMA 1. If $r < s$ and R_k is met at stage r and at stage s, then there is a u such that $r < u < s$ and R_k is injured at stage u.

PROOF. We have $g(r) = g(s) = k$, $F^r \not\subseteq T_{r-1}$ and $F^r \subseteq T_r$. It follows that $H^r \cap T_{s-1} \neq 0$, since otherwise clause (b) would be true at stage s and $R_{g(s)}$ would not be met at stage s. We know that $H^r \cap T_{r-1} = 0$, since otherwise clause (c) would be true at stage r and $R_{g(r)}$ would not be met at stage r. Then $H^r \cap T_r = 0$, since $H^r \cap F^r = 0$ for all r. Thus there is a unique u such that $r < u < s$, $H^r \cap T_{u-1} = 0$ and $H^r \cap T_u \neq 0$. But then R_k is injured at stage u.

LEMMA 2. For each k, the set $\{s|R_k$ is injured at stage $s\}$ has cardinality less than 2^k.

PROOF. By induction on k. Let $k \geq 0$, and suppose that for each $i < k$, the set $\{s|R_i$ is injured at stage $s\}$ has cardinality less than 2^i. Then the set

$$R = \{s|(E_i)(i < k \;\&\; R_i \text{ is injured at stage } s)\}$$

has cardinality less than $2^k - k$. Let

$$S = \{s|(E_i)(i < k \;\&\; R_i \text{ is met at stage } s)\} \quad .$$

With the help of Lemma 1, we proceed to obtain an upper bound on the cardinality of S. Let m_i be the cardinality of the set $\{s|R_i$ is met at stage $s\}$. Then the cardinality of S is at most $\sum_{i=0}^{k-1} m_i$. By Lemma 1, the cardinality of the set $I_i = \{s|R_i$ is injured at stage $s\}$ is at least $m_i - 1$. Since for each $i < k$, I_i has cardinality less than 2^i, is must be that

$$\sum_{i=0}^{k-1} (m_i - 1) < 2^k - k \quad .$$

Thus the cardinality of S is less than 2^k.

We now show that if R_k is injured at stage s, then there is an $i < k$ such that R_i is met at stage s. It will then be clear that the set $\{s|R_k$ is injured at stage $s\}$ has cardinality less than 2^k, since otherwise the cardinality of S would be greater than or equal to 2^k. Fix s and suppose R_k is injured at stage s. Then there is an $r < s$ such that R_k was met at stage r, $H^r \cap T_{s-1} = 0$ and $H^r \cap T_s \neq 0$. Thus we have $0 < r < s$, $F^r \not\subseteq T_{r-1}$, $F^r \subseteq T_r$, $H^r \cap T_{s-1} = 0$ and $H^r \cap F^s \neq 0$. If $g(r)$ were less than or equal to $g(s)$, then either clause (a) or clause (b) would be true at stage s, and T_s would equal T_{s-1}. But $T_{s-1} \neq T_s = T_{s-1} \cup F^s$. Hence $g(s) < g(r) = k$, and $R_{g(s)}$ is met at stage s.

It is an immediate consequence of Lemmas 1 and 2 that for each k, the set $\{s|R_k$ is met at stage $s\}$ has cardinality at most 2^k.

Our Lemma 2 corresponds to Lemma 1 of [1]. In the latter Lemma, Friedberg proved that a certain function $x_1^s(e)$ has the property that for each e, the set $\{x_1^s(e) | s = 0, 1, 2, \ldots\}$ is finite; he could have proved, had he wished to, that the set $\{x_1^s(e) | s = 0, 1, 2, \ldots\}$ has cardinality at most 2^{2e}, but he did not need this additional information. It will be seen in the proof of Theorem 1 below that the only information we need concerning the cardinality of $\{s | R_k$ is injured at stage $s\}$ is that it is finite. We have chosen to prove Lemma 2 in a form stronger than required in order to make a conceptual distinction between the priority method as manifested in [1] and as manifested in Section 5.

In the arguments above we have used the terms <u>requirement</u> and <u>injury</u> in a precise, technical sense. In what follows we surround these terms with quotation marks in order to indicate that they are being used in an intuitive manner suggested by their technical definition. In every classical use of the priority method, there are conflicting "requirements," each one of which is "injured" only finitely often thanks to a happy assignment of priorities.

In Section 5 we will prove several theorems on the degrees of recursively enumerable sets. We will prove a lemma corresponding to Lemma 2 of the present Section, but the "requirements" of Section 5 will be of such a nature that we will merely prove that each "requirement" is "injured" finitely often; we will be unable to prove, as we did in Lemma 2 above, that the kth "requirement" is "injured" less than 2^k times; in fact, we will be unable to produce a recursive function f with the property that the kth requirement is "injured" less than $f(k)$ times.

> THEOREM 1. If t enumerates requirements, then the priority set of t is recursively enumerable in t and meets every t-dense requirement.

PROOF. Let T be the priority set of t. To see that T is recursively enumerable in t, it is sufficient to see that T_s (regarded as a function of s) is recursive in t. But this last is clear from the definition of T.

Fix k and let R_k be t-dense. It is our intention to find an

r such that R_k is met at stage r and is not injured at any stage after stage r. If such an r exists, then $g(r) = k$, $F^r \subseteq T_r \subseteq T$, $H^r \cap T_m = 0$ for all $m \geq r$, and consequently T meets R_k. By Lemmas 1 and 2, there is a u with the following property: if $i \leq k$ and $s \geq u$, then R_i is neither met nor injured at stage s. Let

$$L = \bigcup_{w \leq u} (H^w \cup F^w) \quad .$$

Since R_k is t-dense, there is an $s > 0$ such that $g(s) = k$, $F^s \not\subseteq T_{s-1}$, $H^s \cap T_{s-1} = 0$ and $L \cap F^s = 0$. It is clear that $s > u$, since otherwise $L \cap F^s \neq 0$. This means that R_k is not met at stage s and that consequently $T_{s-1} = T_s \neq T_{s-1} \cup F^s$. Thus either (a) or (b) or (c) is true at stage s. Clause (c) is false at stage s, since $H^s \cap T_{s-1} = 0$. Suppose clause (a) is true at stage s; then there is an $r < s$ such that $g(r) < g(s) = k$, $R_{g(r)}$ is met at stage r and $H^r \cap F^s \neq 0$. But $R_{g(r)}$ is met for the last time prior to stage u because $g(r) < k$; hence $r < u$. But then $H^r \subseteq L$; however, this last is impossible because $H^r \cap F^s \neq 0$ and $L \cap F^s = 0$.

It follows that clause (b) must be true at stage s. Thus there exists an $r < s$ such that $g(r) = g(s) = k$, $r > 0$, $F^r \not\subseteq T_{r-1}$, $F^r \subseteq T_r$ and $H^r \cap T_{s-1} = 0$. If there were an m such that $H^r \cap T_m \neq 0$, then m would be greater than or equal to s, hence greater than u, and R_k would be injured at some stage after stage u, which is impossible. Consequently $F^r \subseteq T$, $H^r \cap T = 0$ and T meets R_k.

It is not hard to see that there is some connection between Theorem 1 above and Baire's category theorem. This matter will be discussed further in Section 10, where we will consider the application of category and measure-theoretic methods to the problem of existence proofs for degrees. We will also have more to say about the priority method in Sections 6 and 9.

We have presented Theorem 1 in such a way as to suggest that an application of it consists of the definition of a function t which enumerates requirements followed by an assertion concerning the properties of T, the priority set of t. In actuality, Theorem 1 is applied in the form of a simultaneous, inductive definition of t and T. This is possible, since for each s, the value of T_s was defined above solely in terms of

$t(0), t(1), \ldots, t(s), T_0, T_1, \ldots, T_{s-1}$.

COROLLARY 1. (Friedberg [1], Muchnik [13]) There exist two recursively enumerable sets whose degrees of unsolvability are incomparable.

PROOF. We define functions T_s, F^s, H^s and $g(s)$ simultaneously by induction.

Stage $s = 0$. We set $T_0 = F^0 = H^0 = 0$ and $g(0) = 0$.

Stage $s > 0$. Let $A_s = \{n \mid 2n \in T_{s-1}\}$ and let

$$B_s = \{n \mid 2n + 1 \in T_{s-1}\}.$$

Let f_s^0 and f_s^1 be the representing functions of A_s and B_s respectively. Let $i = 1 \dot{-} (s)_0$ and $e = (s)_1$.

CASE 1. $(Em)_{0 \leq m \leq s}(Ey)_{y \leq s}\left[T_1^1\left(f_s^i(y), e, p_e^m, y\right) \;\&\; U(y) = 1\right]$.

Let r be the greatest $m \leq s$ such that

$$(Ey)_{y \leq s}\left[T_1^1\left(f_s^i(y), e, p_e^m, y\right) \;\&\; U(y) = 1\right] .$$

We set

$$F^s = \{2p_e^r + 1 - i\} \quad ;$$

$$H^s = \{2n + i \mid f_s^i(n) = 1 \;\&\; n \leq s\} \quad ;$$

$$g(s) = 2e + i + 1 \quad .$$

CASE 2. Otherwise. We set

$$F^s = H^s = 0 \quad \text{and} \quad g(s) = 0 \quad .$$

In both cases we define T_s as in the proof of Theorem 1.

For all $s \geq 0$, we set

$$t(s) = 2^{j(F^s)} \cdot 3^{j(H^s)} \cdot 5^{g(s)} \quad .$$

It is clear that t enumerates requirements and that $T = \bigcup_{s=0}^{\infty} T_s$ is the priority set of t. Note that if Case 2 holds at stage $s > 0$, then $T_s = T_{s-1}$, since $F^s = 0$. Let $A = \{n \mid 2n \in T\}$ and $B = \{n \mid 2n + 1 \in T\}$. Then $A = \bigcup_{s>0} A_s$ and $B = \bigcup_{s>0} B_s$. Let f^0 and f^1 be the representing functions of A and B respectively. It follows that for each $i < 2$ and each n, $f^i(n) = f_s^i(n)$ for all sufficiently large s. Since only bounded quantifiers occur in the hypothesis of Case 1, the functions t

and T_s must be recursive. Thus the sets T, A and B are recursively enumerable. By Theorem 1, T meets every t-dense requirement; we use this fact to show B is not recursive in A with Gödel number e.

We suppose that $\{e\}^{f^0}$ is the representing function of a set of natural numbers, since otherwise there is nothing to prove. We first suppose that R_{2e+1} is t-dense, and then suppose otherwise. Suppose then that R_{2e+1} is t-dense. By Theorem 1, T meets R_{2e+1}. This means there is an s such that $g(s) = 2e + 1$, $F^s \subseteq T$ and $H^s \cap T = 0$. Then s must be such that $i = 1 \dot{-} (s)_0 = 0$, $e = (s)_1$ and Case 1 is true at stage s. But then there is an r and a y such that

$$y \leq s \;\&\; T_1^1\left(\tilde{f}_s^0(y),\ e,\ p_e^r,\ y\right) \;\&\; U(y) = 1 \quad ;$$
$$F^s = \{2p_e^r + 1\} \quad ;$$
$$H^s = \{2n \mid f_s^0(n) = 1 \;\&\; n \leq s\} \quad .$$

Since $F^s \subseteq T$, we have $p_e^r \in B$ and $f^1(p_e^r) = 0$. Since $H^s \cap T = 0$, we have $f^0(n) = f_s^0(n)$ for all $n \leq s$. Since $y \leq s$, it must be that

$$T_1^1\left(\tilde{f}^0(y),\ e,\ p_e^r,\ y\right) \;\&\; U(y) = 1 \quad .$$

Then $\{e\}^{f^0}(p_e^r) = 1$, and B is not recursive in A with Gödel number e.

Now suppose that R_{2e+1} is not t-dense. Note that the functions H^s and T_s have been defined in such a manner that $H^s \cap T_{s-1} = 0$ for all s. It follows that there is a finite set L such that for all $s > 0$, either $g(s) \neq 2e + 1$ or $F^s \subseteq T_{s-1}$ or $L \cap F^s \neq 0$. We will use this last fact to prove: if $m > 0$ and $2p_e^m + 1$ is greater than every member of L, then (i) $p_e^m \notin B$ and (ii) $\{e\}^{f^0}(p_e^m) = 0$. It will then be clear that B is not recursive in A with Gödel number e. Let m be such that $m > 0$ and $2p_e^m + 1$ is greater than every member of L. Suppose $p_e^m \in B$. Then $2p_e^m + 1 \in T$. Let $s > 0$ be such that $2p_e^m + 1 \in T_s - T_{s-1}$. Then $F^s = \{2p_e^m + 1\}$ and $T_s = T_{s-1} \cup F^s$. Thus $g(s) = 2e + 1$, $F^s \not\subseteq T_{s-1}$ and $L \cap F^s = 0$. Since this last is impossible, (i) is proved.

Suppose $\{e\}^{f^0}(p_e^m) = 1$. Then there is a y such that

$$T_1^1\left(\tilde{f}^0(y),\ e,\ p_e^m,\ y\right) \;\&\; U(y) = 1 \quad .$$

Since $f^0(y) = f_s^0(y)$ for all sufficiently large s, it follows that there is an s such that $i = 1 \dot{-} (s)_0 = 0$, $e = (s)_1$, $m \leq s$, $y \leq s$ and

Case 1 is true at stage s. Then $F^s = \{2p_e^r + 1\}$, where $r \geq m$, and $g(s) = 2e + 1$. Since $r \geq m$, we have $F^s \not\subseteq T_{s-1}$ by (i). Thus $g(s) = 2e + 1$, $F^s \not\subseteq T_{s-1}$ and $L \cap F^s = 0$. Since this last is impossible, (ii) is proved.

That completes the proof that B is not recursive in A. Only notational changes need be made in the above argument to show A is not recursive in B.

Let $\lambda mn|A(m, n)$ be a function such that for each $m \geq 0$, $\lambda n|A(m, n)$ is the representing function of a set A_m of natural numbers. Following Kleene and Post [9], we say the sets $A_0, A_1, A_2, \ldots,$ are recursively independent (in sequence) if for each $u \geq 0$, the set A_u is not recursive in the function $\lambda mn|A(m + sg((m + 1) \dot{-} u), n)$. The result expressed in Corollary 2 was announced without proof by Muchnik in [14].

COROLLARY 2. There exists a sequence of recursively independent, simultaneously recursively enumerable sets.

PROOF. Our construction is virtually indistinguishable from that of Corollary 1. Again we define functions T_s, F^s, H^s and $g(s)$ simultaneously by induction.

Stage s = 0. We set $T_0 = F^0 = H^0 = 0$ and $g(0) = 0$.

Stage s > 0. For each $m \geq 0$, let $\lambda n|A^s(m, n)$ be the representing function of the set $\{n|p_m^{1+n} \in T_{s-1}\}$. Let $u = (s)_0$ and $e = (s)_1$. Let f_s^u denote the function

$$\lambda mn|A^s(m + sg((m + 1) \dot{-} u), n) \quad .$$

CASE 1. $(Em)_{0 < m \leq s}(Ey)_{y \leq s}\left[T_1^2\left(\tilde{f}_s^u(y, y), e, p_e^m, y\right) \,\&\, U(y) = 1\right]$.

Let r be the greatest $m \leq s$ such that

$$(Ey)_{y < s}\left[T_1^2\left(\tilde{f}_s^u(y, y), e, p_e^m, y\right) \,\&\, U(y) = 1\right] \quad .$$

We set

$$F^s = \left\{p_u^{1+p_e^r}\right\} \quad ;$$

$$H^s = \left\{p_{m+sg((m+1) \dot{-} u)}^{1+n} \,\middle|\, f_s^u(m, n) = 1 \,\&\, m \leq s \,\&\, n \leq s\right\} \quad ;$$

$$g(s) = 2^u \cdot 3^e \quad .$$

CASE 2. Otherwise. We set

$$F^s = H^s = 0 \quad \text{and} \quad g(s) = 0.$$

In both cases we define T_s as in the proof of Theorem 1.

For all $s \geq 0$, we set

$$t(s) = 2^{j(F^s)} \cdot 3^{j(H^s)} \cdot 5^{g(s)} \quad .$$

It is clear that t enumerates requirements and that $T = \bigcup_{s=0}^{\infty} T_s$ is the priority set of t. For each $u \geq 0$, let $A_u = \{n | p_u^{1+n} \in T\}$, and let $\lambda n | A(u, n)$ be the representing function of A_u. Since T is recursively enumerable, the sets $A_0, A_1, A_2, \ldots,$ are simultaneously recursively enumerable. We claim that the sets $A_0, A_1, A_2, \ldots,$ are recursively independent (in sequence). We merely sketch the proof since it is so similar to that of Corollary 1.

Let f^u denote the function $\lambda mn | A(m + sg((m+1) \dot{-} u), n)$. We show that A_u is not recursive with Gödel number e in f^u. Suppose that $\{e\}^{f^u}$ is the representing function of a set of natural numbers, since otherwise there is nothing to prove. Let $w = 2^u \cdot 3^e$. We proceed as in Corollary 1. If R_w is t-dense, then T meets R_w by Theorem 1, and there is an s such that $g(s) = w$, $F^s \subseteq T$ and $H^s \cap T = 0$. Then $F_s = \{p_u^{1+p_e^r}\}$ and $p_e^r \in A_u$. But $\{e\}^{f^u}(p_e^r) = \{e\}^{f_s^u}(p_e^r) = 1$, since $\tilde{f}_s^u(s, s) = \tilde{f}^u(s, s)$ as a consequence of $H^s \cap T = 0$.

If R_w is not t-dense, then there is a finite set L such that for all $s > 0$, either $g(s) \neq w$ or $F^s \subseteq T_{s-1}$ or $L \cap F^s \neq 0$. If follows that if $m > 0$ and $p_u^{1+p_e^m}$ is greater than every member of L, then $p_e^m \notin A_u$ and $\{e\}^{f^u}(p_e^m) = 0$.

In [9] Kleene and Post constructed a sequence of recursively independent sets of degree less than $0'$, and then, with the help of this sequence, showed that the rationals (with the usual ordering) could be imbedded in the upper semi-lattice of degrees less than or equal to $0'$. We extend the idea underlying their argument in order to prove the next corollary.

COROLLARY 3. If P is a countable, partially ordered set, then P is imbeddable in the upper semi-lattice of recursively enumerable degrees.

PROOF. Let N be the set of all natural numbers, and let $\leq_R$ be a partial ordering relation for N. We say $\leq_R$ is recursive if the predicate $m \leq_R n$ is recursive, and we denote by N_R the partially ordered set whose range is N and whose relation is $\leq_R$. Mostowski [12] announced the following result in somewhat different form: there exists a recursive, partial ordering relation $\leq_R$ such that every countable, partially ordered set is imbeddable in N_R. Since P is imbeddable in N_R, we need only show N_R is imbeddable in the upper semi-lattice of degrees of recursively enumerable sets.

Let $A_0, A_1, A_2, \ldots,$ be a sequence of recursively independent, simultaneously recursively enumerable sets. For each $m \geq 0$, let $B_m = \{p_m^{n+1} | n \in A_m\}$. Then $B_0, B_1, B_2, \ldots,$ is a sequence of recursively independent, disjoint, simultaneously recursively enumerable sets. For each $m \geq 0$, let $\lambda n | B(m, n)$ be the representing function of B_m, and let $C_m = \cup \{B_r | r \leq_R m\}$; C_m is recursively enumerable, since the B_m's are simultaneously recursively enumerable, and since $\leq_R$ is recursive. (In fact the C_m's are simultaneously recursively enumerable.) We will now show that C_u is recursive in C_v if and only if $u \leq_R v$. It will then be clear that N_R is imbeddable in the upper semi-lattice of degrees of recursively enumerable sets.

Suppose $u \not\leq_R v$. Then $B_u \cap C_v = 0$, and C_v is recursive in $\lambda mn | B(m + sg((m+1) \dot{-} u), n)$, since

$$x \in C_v \leftrightarrow (Em)(En)\left(x = p_m^{n+1} \ \&\ m \leq_R v \ \&\ x \in B_m\right) \quad .$$

It follows from the recursive independence of $B_0, B_1, B_2, \ldots,$ that B_u is not recursive in C_v. But then C_u is not recursive in C_v, since B_u is recursive in C_u.

Suppose $u \leq_R v$. Then $C_u \subseteq C_v$, and C_u is recursive in C_v, since

$$\begin{aligned} x \in C_u &\leftrightarrow (Em)(En)\left(x = p_m^{n+1} \ \&\ m \leq_R u \ \&\ x \in B_m\right) \\ &\leftrightarrow (Em)(En)\left(x = p_m^{n+1} \ \&\ m \leq_R u \ \&\ x \in C_v\right) \quad . \end{aligned}$$

This last is a consequence of the fact that the B_m's are disjoint.

The proof of Corollary 3 provides us with a simple approach to the problem of imbedding an arbitrary, countable, partially ordered set in a given set of degrees. We will make use of this approach in Section 5 to prove: if $\underline{b}$ and $\underline{c}$ are degrees such that $\underline{b} < \underline{c}$ and $\underline{c}$ is recursively enumerable in $\underline{b}$, and if S is the upper semi-lattice of all degrees greater than $\underline{b}$, less than or equal to $\underline{c}$ and recursively enumerable in $\underline{b}$, then every countable, partially ordered set is imbeddable in S.

§5. AN EXISTENCE THEOREM FOR RECURSIVELY ENUMERABLE DEGREES

A set D is said to be recursively enumerable in a set B if there is a function recursive in B whose range is D.

In Section 4 we saw that there exist two incomparable, recursively enumerable degrees. In the present section we will see that given any non-zero degree less than $\underline{0}'$, there exists a recursively enumerable degree imcomparable with it. We will also see that the priority method as manifested below is less constructive than as manifested in Section 4.

THEOREM 1. Let B, C and D be sets such that C is recursive in D, D is recursively enumerable in B, and C is not recursive in B. Then D is the disjoint union of sets D_0 and D_1 such that for each $i < 2$, D_i is recursively enumerable in B, and C is not recursive in B, D_i.

PROOF. We first show it is safe to assume every member of B is even and every member of D is odd. Let

$$B^* = \{2n \mid n \in B\} \quad \text{and}$$
$$D^* = \{2n+1 \mid n \in D\} \quad .$$

Then C is recursive in D^*, D^* is recursively enumerable in B^*, and C is not recursive in B^*. Suppose there are disjoint sets D_0^* and D_1^* such that $D_0^* \cup D_1^* = D^*$ and such that for each $i < \underline{2,}$ D_i^* is recursively enumerable in B^*, and C is not recursive in B^*, D_i^*. Then the disjoint sets

$$D_0 = \{n \mid 2n+1 \in D_0^*\} \quad \text{and}$$
$$D_1 = \{n \mid 2n+1 \in D_1^*\}$$

are such that $D_0 \cup D_1 = D$ and such that for each $i < 2$, D_i is recursively enumerable in B, and C is not recursive in B, D_i.

Let b, c and d be the representing functions of B, C and D respectively. Then $b \cdot d$ is the representing function of $B \cup D$. D is infinite, because there is a non-recursive set, namely C, which is recursive in D. Let f be a function recursive in B which enumerates D without repetitions:

$$D = \{f(n) \mid n \geq 0\} \quad \text{and}$$
$$(n)(m)(n \neq m \rightarrow f(n) \neq f(m)) \quad .$$

Let

$$d(s, n) = \begin{cases} 0 & \text{if } (Ek)(k \leq s \ \& \ f(k) = n) \\ 1 & \text{otherwise .} \end{cases}$$

It is clear that the function d(s, n) is recursive in B and that for each m, $\lim_s d(s, n)$ exists and is equal to d(n). Since C is recursive in D, there exists a Gödel number g with the property that $c(n) = \{g\}^d(n)$ for all n. We make use of g to define a function c(s, n):

$$c(s, n) = \begin{cases} U\Big(\mu y T_1^1\big(\prod_{j<y} p_j^{d(s,j)}, g, n, y\big)\Big) & \text{if} \\ \qquad (Ey)\Big(y \leq s \ \& \ T_1^1\big(\prod_{j<y} p_j^{d(s,j)}, g, n, y\big)\Big) & \\ s + 1 & \text{otherwise.} \end{cases}$$

It is clear that the function c(s, n) is recursive in B. Since $\lim_s d(s, n) = d(n)$ for all n, and since $c(n) = \{g\}^d(n)$ for all n, it follows that $\lim_s c(s, n)$ exists and is equal to c(n) for all n.

A very convenient property of the Gödel numbering devised by Kleene in [7] to arithmetize his formalism for recursive functions is: the Gödel number of a deduction is greater than the intuitive counterpart of any formal numeral occurring in the deduction. We will denote this fact by GND. Friedberg utilized GND in [1]. It follows from GND that $c(s, n) = s+1$ whenever $n \geq s$, and that $c(s, n) \leq s+1$ for all n and s.

It is our intention to define twelve functions simultaneously by induction: $d_i(s, n)$, $t_i(s)$, $y_i(s, n, e)$, $P_i(s, n, e)$, $m_i(s, e)$ and $K_i(s, e)$ $(i = 0, 1)$. Each of these functions will be recursive in B. At stage $s (s \geq 0)$ of the induction, we will define $d_i(s, n)$, $t_i(s)$, $y_i(s, n, e)$, $P_i(s, n, e)$ $m_i(s, e)$ and $K_i(s, e)$ for all $i < 2$, all n and all e. The sole purpose of stage s is to put the natural number

$f(s)$ (the sth member of D) in D_0 or in D_1, but not in both. Thus at stage s we set either $d_0(s, f(s))$ or $d_1(s, f(s))$ equal to 0, but not both.

Stage $s = 0$. We set $d_1(0, n) = 1$ for all $n \neq f(0)$, and we set $d_1(0, f(0)) = 0$. We set $P_i(0, n, e) = 2$, $d_0(0, n) = t_i(0) = y_i(0, n, e) = 1$ and $m_i(0, e) = K_i(0, e) = 0$ for all $i < 2$, all e and all n.

Stage $s > 0$. For each $i < 2$, let

$$t_i(s) = \mu e_{e \leq s}[f(s) < K_i(s-1, e)] \quad .$$

Recall that the bounded least number operator is defined in such a way that $t_i(s) = s + 1$ if and only if

$$\sim (Ee)_{e \leq s}[f(s) < K_i(s-1, e)] \quad .$$

Let $z(s) = 1$ if $t_1(s) \geq t_0(s)$, and let $z(s) = 0$ otherwise. Thus $t_{z(s)}(s) \geq t_{1-z(s)}(s)$. We set

$$d_i(s, n) = \begin{cases} 0 & \text{if } i = z(s) \,\&\, n = f(s) \\ d_i(s-1, n) & \text{otherwise} \end{cases}$$

for all $i < 2$ and all n.

We define $y_i(s, n, e)$ and $P_i(s, n, e)$ for all $i < 2$ and all n and e:

$$y_i(s, n, e) = \mu y_{y \leq s}\left[T_1^1\Big(\prod_{j < y} p_j^{b(j) \cdot d_i(s,j)}, e, n, y\Big)\right] \quad ;$$

$$P_i(s, n, e) = \begin{cases} U\big(y_i(s, n, e)\big) & \text{if } y_i(s, n, e) \leq s \\ s + 2 & \text{otherwise.} \end{cases}$$

Before we define $m_i(s, e)$, we note that

$$(i)_{i \leq 2}(e)(Et)_{t \leq s}[c(s, t) \neq P_i(s, t, e)] \quad .$$

This last is clear, since it is a consequence of GND that $c(s, s) = s + 1$ and that $P_0(s, s, e) = P_1(s, s, e) = s + 2$ for all e. We complete stage $s > 0$ by defining $m_i(s, e)$ and $K_i(s, e)$ for all $i < 2$ and all e:

$$m_i(s, e) = \mu t[c(s, t) \neq P_i(s, t, e)] \quad ;$$

$$K_i(s, e) = \begin{cases} K_i(s-1, e) \text{ if } (n)[n < m_i(s, e) \rightarrow y_i(s, n, e) \leq K_i(s-1, e)] \\ \mu t(n)[n < m_i(s, e) \rightarrow y_i(s, n, e) \leq t] \quad \text{otherwise.} \end{cases}$$

We list two remarks which will be needed in Lemmas 2 and 4:

(R1) $(i)_{i \leq 2}(s)(e)[K_i(s+1, e) \geq K_i(s, e)]$;

(R2) $(i)_{i \leq 2}(s)(e)(n)[n < m_i(s, e) \rightarrow y_i(s, n, e) \leq K_i(s, e)]$

Both remarks are immediate from the definition of K_i.

We use the method described in Section 1 to show that each of the twelve functions we have defined is recursive in B. For each $i < 2$, let $Q_i(s, n, e)$ denote

$$2^{t_i(s)} \cdot 3^{d_i(s,n)} \cdot 5^{y_i(s,n,e)} \cdot 7^{P_i(s,n,e)} \cdot 11^{m_i(s,e)} \cdot 13^{K_i(s,e)}$$

Let

$$Q(s, n, e) = 2^{Q_0(s,n,e)} \cdot 3^{Q_1(s,n,e)}$$

It will be sufficient to show that the function Q is recursive. For this purpose we need a function H such that

$$Q(s, n, e) = H\big(s-1, Q(s-1, n, e), n, e\big)$$

for all $s > 0$, and such that H is recursive in B. We do not actually exhibit H, but the following list of equations makes clear that such an H exists; recall that each of the functions $f(s)$, $b(j)$ and $c(s, n)$ is recursive in B; let $i = 0, 1$.

$$(Q(s, n, e))_{i,0} = \mu e_{e \leq s}\Big[f(s) < \big(Q(s-1, n, e)\big)_{i,5}\Big] ,$$

$$(Q(s, n, e))_{i,1} = \begin{cases} 0 & \text{if } n = f(s) \;\&\; (Q(s, n, e))_{i,0} > (Q(s, n, e))_{1-i,0} \\ 0 & \text{if } n = f(s) \;\&\; 1 = i \;\&\; (Q(s, n, e))_{i,0} = \\ & \qquad (Q(s, n, e))_{1-i,0} \\ (Q(s-1, n, e))_{i,1} & \text{otherwise,} \end{cases}$$

$$(Q(s, n, e))_{i,2} = \mu y_{y \leq s}\Big[T_1^1\Big(\prod_{j<y} p_j^{\,b(j) \cdot (Q(s, j, e))_{i,1}}, e, n, y\Big)\Big]$$

$$(Q(s, n, e))_{i,3} = \begin{cases} U\big((Q(s, n, e))_{i,2}\big) & \text{if } \big(Q(s, n, e)\big)_{i,2} \leq s \\ s + 2 & \text{otherwise,} \end{cases}$$

$$(Q(s, n, e))_{i,4} = \mu t_{t \leq s}\Big[c(s, t) \neq \big(Q(s, t, e)\big)_{i,3}\Big] ,$$

$$\big(Q(s, n, e)\big)_{i,5} = \begin{cases} \big(Q(s-1, n, e)\big)_{i,5} \text{ if } (n)\Big[n < \big(Q(s, n, e)\big)_{i,4} \\ \qquad \rightarrow \big(Q(s, n, e)\big)_{i,2} \leq \big(Q(s-1, n, e)\big)_{i,5}\Big] \\ \mu t(n)\Big[n < \big(Q(s, n, e)\big)_{i,4} \rightarrow \big(Q(s, n, e)\big)_{i,2} \leq t\Big] \\ \qquad\qquad \text{otherwise.} \end{cases}$$

Before we proceed with the rigorous, mathematical argument, let us consider the intuitive content of the above construction. We have that C is not recursive in B. We wish to define sets D_0 and D_1 such that C is not recursive in B, D_i for any $i < 2$. In addition we require that each of the sets D_0 and D_1 be recursively enumerable in B and that D be the disjoint union of D_0 and D_1. Since we have that D is recursively enumerable in B, it is only reasonable that our construction should have the following form: we enumerate the members of D without repetitions by means of a function recursive in B; at stage s we enumerate the sth member of D and deposit it in just one of the sets D_0 and D_1; we choose between D_0 and D_1 according to a criterion based solely on our desire that C not be recursive in B, D_i for any $i < 2$. The functions t, y_i, P_i, m_i and K_i serve to establish this criterion. The value of $z(s)$ at $s > 0$ is 0 or 1 according to whether we put the sth member of D in D_0 or D_1.

In order that D_i be recursively enumerable in B, the above functions must be recursive in B. Two obstacles separate us from this objective. First, C is not recursive in B. Thus at stage s, when we try to make some estimate of our progress towards our goal of defining D_i in such a manner that C is not recursive in B, D_i, we must make this estimate without any perfect knowledge of the membership of C. However, we have that C is recursive in D and that D is recursively enumerable in B. This last makes it possible to define an approximation of the representing function of C at stage s. The function $c(s, n)$ is recursive in B, and $\lim_s c(s, n)$ is the representing function of C.

The second obstacle consists of our inability to know at stage s whether or not we have "met" some "requirement" once and for all. (The terms surrounded by quotation marks are being used in the intuitive sense of Section 4.) Our "requirements" are: C is not recursive in B, D_i with Gödel number e, where $i < 2$ and $e \geq 0$. It may appear that a "requirement" is "met" at stage s, but then at a later stage, a change in the membership of D_i or in the approximation of the membership of C may alter things completely. Thus we find ourselves in a situation similar to that of Section 4. There are conflicting "requirements" that must be "met," and we are forced to make repeated attempts to "meet" them with the hope that eventually each of them will be "met."

For each $i < 2$ and $e \geq 0$, let R_{2e+i} denote the "requirement" that C not be recursive in B, D_i with Gödel number e. Our system of priorities is the same as that of Section 4; that is, R_n has higher priority than R_m if $n < m$. At stage s we examine all deductions whose Gödel numbers are less than or equal to s in order to determine as far as we can at stage s the result of applying the eth recursive, reduction procedure to B, D_i. The function $P_i(s, n, e)$ (regarded as a function of n) is merely an approximation of $\{e\}^{b,d_i}$ at stage s. We compare this approximation with $c(s, n)$ in order to define an initial segment of the natural numbers on which the functions $\{e\}^{b,d_i}$ and c appear to agree as of stage s; the length of the initial segment is $m_i(s, e)$. The value of $K_i(s, e)$ is greater than or equal to the Gödel numbers of the deductions needed to establish the apparent equality of $\{e\}^{b,d_i}(n)$ and $c(n)$ for all $n < m_i(s, e)$.

When we choose between D_0 and D_1 at stage s, our choice is motivated by a desire to preserve as far as is possible the apparent equality between $\{e\}^{b,d_i}$ and c noted at stage s-1. If $f(s) \geq K_i(s-1, e)$, then the apparent equality between $\{e\}^{b,d_i}(n)$ and $c(n)$ for all $n < m_i(s-1, e)$ will not be disturbed if $f(s)$ is added to D_i. This is so because $f(s)$ is greater than the Gödel number of any deduction relevant to the apparent equality, and hence by GND, the addition of $f(s)$ to D_i will not affect any such deduction. If $f(s) < K_i(s-1, e)$, then we arbitrarily regard the addition of $f(s)$ to D_i at stage s as an "injury" to "requirement" R_{2e+i}. The value of $t_i(s)$ is the least e less than or equal to s such that R_{2e+i} will be "injured" if $f(s)$ is put in D_i;

if no such e exists, $t_1(s) = s + 1$. Thus $R_{2t_1(s)+1}$ is the highest priority "requirement" that will be "injured" (if any) when $f(s)$ is added to D_1. The function $z(s)$ is defined in such a manner that when we are faced with the alternatives of "injuring" $R_{2t_0(s)}$ or $R_{2t_1(s)+1}$, we choose to "injure" the "requirement" of lower priority.

It follows that we will never "injure" R_0, that we will "injure" R_1 only to avoid "injuring" R_0, and that in general we will "injure" R_m at stage s only to avoid "injuring" R_n for some $n < m$. R_{2e+i} will be "met" if the set $\{K_i(s, e) | s \geq 0\}$ is finite, because if the latter set is finite, then $c(n)$ and $\{e\}^{b,d_i}(n)$ will appear to be equal for only finitely many n during the course of the construction, and consequently at the end of the construction there will be an n such that either $\{e\}^{b,d_i}(n)$ is undefined or it is not equal to $c(n)$. We will show by means of a simultaneous induction (in the form of an infinite descent) that for each $i < 2$ and each e:

(a) R_{2e+i} is "injured" only finitely often;

(b) the set $\{K_i(s, e) | s \geq 0\}$ is finite (R_{2e+i} is "met").

In Section 4 we first showed that each "requirement" is "injured" only finitely often and then showed that each "requirement" is "met"; that is, we separated the proofs of statements analogous to (a) and (b). In the present section we are unable to separate the proofs of (a) and (b). This is the reason that we are unable in the present section to provide an effective bound on the number of times each "requirement" is "injured" in the course of the construction. Recall that in Section 4 we proved R_n is "injured" less than 2^n times. We could, if we wished to, define a function h such that each "requirement" R_n of the present section is "injured" less than $h(n)$ times and such that h has degree less than or equal to $\underline{0}'$.

For each $i < 2$ and each n, let

$$d_i(n) = \lim_s d_i(s, n) \quad ,$$

and let D_i be the set whose representing function is d_i. Then $D_0 \cap D_1 = 0$ and $D = D_0 \cup D_1$, since at stage s, either $d_0(s, f(n))$ or $d_1(s, f(n))$ was set equal to 0, but not both.

For the sake of a reductio ad absurdum we suppose there is an $i < 2$ and an e such that (b) is false. Let e^* be the least e such that the set $\{K_i(s, e) | i < 2 \& s \geq 0\}$ is infinite. Let i^* be the least i such that $\{K_i(s, e^*) | s \geq 0\}$ is infinite. Our objective is to show that C is recursive in B, contrary to the hypothesis of our theorem. First we show that $R_{2e^*+i^*}$ is "injured" only finitely often.

LEMMA 1. There is an s' such that for all $s \geq s'$, either $z(s) = 1-i^*$ or $e^* < t_{i^*}(s)$.

PROOF. Suppose there are infinitely many s such that $z(s) = i^*$ and $e^* \geq t_{i^*}(s)$. Let S be an infinite set such that for all $s \in S$, $s \geq e^*$, $z(s) = i^*$ and $e^* \geq t_{i^*}(s)$. Recall that $t_{z(s)}(s) \geq t_{1-z(s)}(s)$ for all $s > 0$. Then for each $s \in S$,

$$e^* \geq t_{i^*}(s) \geq t_{1-i^*}(s) \geq 0 \quad .$$

Since S is infinite, there must be an infinite subset R of S and an e^{**} such that $t_{1-i^*}(s) = e^{**} \geq 0$ for all $s \in R$. Thus we have

$$s \geq e^* \geq t_{i^*}(s) \geq t_{1-i^*}(s) = e^{**}$$

for all $s \in R$. It follows that $f(s) < K_{1-i^*}(s-1, e^{**})$ for all $s \in R$. But then the set $\{K_{1-i^*}(s-1, e^{**}) | s \in R\}$ is infinite, since the set $\{f(s) | s \in R\}$ is infinite. (Recall that f is a one-one function.) It follows from the definition of e^* that $e^* = e^{**}$. Thus the set $\{K_{1-i^*}(s, e^*) | s \geq 0\}$ is infinite. This means that $1-i^* > i^*$, and that consequently $i^* = 0$. But then for any $s \in R \subseteq S$, we have

$$e^* = t_{i^*}(s) = t_{1-i^*}(s) = e^{**}$$

and $z(s) = i^* = 0$. This last is impossible, since $z(s) = 1$ if $t_1(s) = t_0(s)$ and $s > 0$.

Let s^* be the least s' such that $s' > e^*$ and such that for all $s \geq s'$, $z(s) = 1-i^*$ or $e^* < t_{i^*}(s)$.

LEMMA 2. If $s \geq s^*$ and $n < m_{i^*}(s, e^*)$, then $d_{i^*}(s, j) = d_{i^*}(s', j)$ for all j and s' such that $j < y_{i^*}(s, n, e^*)$ and $s' \geq s$.

PROOF. Let s and n be such that $s \geq s^*$ and $n < m_{i*}(s, e^*)$. We prove the lemma by induction on $s' \geq s$. Our induction hypothesis is $s' \geq s$ and $d_{i*}(s, j) = d_{i*}(s', j)$ for all $j < y_{i*}(s, n, e^*)$. Since $s' \geq s^*$, we have that either

$$z(s'+1) = 1 - i^* \quad \text{or} \quad e^* < t_{i*}(s'+1)$$

If $z(s'+1) = 1 - i^*$, then $d_{i*}(s'+1, j) = d_{i*}(s', j)$ for all j. Suppose then that $e^* < t_{i*}(s'+1)$. Then

$$t_{i*}(s'+1) = \mu e_{e \leq s'+1}[f(s'+1) < K_{i*}(s', e)] > e^*$$

It follows that $f(s'+1) \geq K_{i*}(s', e^*)$, since $e^* < s^* < s'+1$. By remark (R1), $K_{i*}(s', e^*) \geq K_{i*}(s, e^*)$, since $s' \geq s$. By remark (R2), $K_{i*}(s, e^*) \geq y_{i*}(s, n, e^*)$, since $n < m_{i*}(s, e^*)$. But then

$$f(s'+1) \geq y_{i*}(s, n, e^*) \quad ,$$

and consequently, $d_{i*}(s'+1, j) = d_{i*}(s', j)$ for all $j < y_{i*}(s, n, e^*)$.

LEMMA 3. If $s \geq s^*$ and $n < m_{i*}(s, e^*)$, then $y_{i*}(s, n, e^*) = y_{i*}(s', n, e^*) \leq s$ for all $s' \geq s$.

PROOF. Let s, n and s' be such that $s' \geq s \geq s^*$ and $n < m_{i*}(s, e^*)$. Clearly $c(s, n) = P_{i*}(s, n, e^*)$. Since $c(s, n) \leq s+1$, we have

$$P_{i*}(s, n, e^*) = U\big(y_{i*}(s, n, e^*)\big)$$

and $y_{i*}(s, n, e^*) \leq s$. But then

$$y_{i*}(s, n, e^*) = \mu y T_1^1\Big(\prod_{j<y} p_j^{b(j) \cdot d_{i*}(s,j)}, e^*, n, y\Big) \quad .$$

By Lemma 2, $d_{i*}(s, j) = d_{i*}(s', j)$ for all $j < y_{i*}(s, n, e^*)$. It follows that $y_{i*}(s, n, e^*) = y_{i*}(s', n, e^*)$.

LEMMA 4. The set $\{m_{i*}(s, e^*) \mid s \geq 0\}$ is infinite.

PROOF. Suppose that the set $\{m_{i*}(s, e^*) \mid s \geq 0\}$ is finite; let m' be its greatest member. Let s' be such that $s' \geq s^*$ and $c(s, n) = \lim_s c(s, n) = c(n)$ for all s and n such that $s \geq s'$ and $n \leq m'$. Let m'' be the greatest member of $\{m_{i*}(s, e^*) \mid s \geq s'\}$. Let s'' be such that $s'' \geq s'$ and $m_{i*}(s'', e^*) = m'' \leq m'$.

We now show by induction on $s \geq s''$ that $m_{i*}(s, e^*) = m''$ and

$K_{i*}(s, e^*) = K_{i*}(s'', e^*)$ for all $s \geq s''$. Suppose then that $s \geq s''$, $m_{i*}(s, e^*) = m''$ and $K_{i*}(s, e^*) = K_{i*}(s'', e^*)$. By Lemma 3,

$$y_{i*}(s, n, e^*) = y_{i*}(s+1, n, e^*) \leq s$$

for all $n < m_{i*}(s, n, e^*)$, since $s \geq s'' \geq s' \geq s^*$. This means that

$$P_{i*}(s, n, e^*) = P_{i*}(s+1, n, e^*) = c(s, n)$$

for all $n < m_{i*}(s, e^*)$. But $c(s+1, n) = c(s, n) = c(n)$ for all $n < m'' \leq m'$, since $s \geq s'$. Hence

$$P_{i*}(s+1, n, e^*) = c(s+1, n)$$

for all $n < m'' = m_{i*}(s, e^*)$, and consequently, $m'' \leq m_{i*}(s+1, e^*)$. It follows from the definition of m'' that $m_{i*}(s+1, e^*) = m''$. But then by remark (R2),

$$y_{i*}(s, n, e^*) = y_{i*}(s+1, n, e^*) \leq K_{i*}(s, e^*)$$

for all $n < m_{i*}(s+1, e^*) = m'' = m_{i*}(s, e^*)$. From this last and the definition of K_{i*}, it is clear that $K_{i*}(s+1, e^*) = K_{i*}(s, e^*)$.

Thus we have shown $K_{i*}(s, e^*) = K_{i*}(s'', e^*)$ for all $s \geq s''$; but this is absurd, since by definition of e^* and i^*, the set $\{K_{i*}(s, e^*) \mid s \geq 0\}$ is infinite.

LEMMA 5. If $s \geq s^*$ and $n < m_{i*}(s, e^*)$, then $c(n) = P_{i*}(s, n, e^*)$.

PROOF. Let $s \geq s^*$ and $n < m_{i*}(s, e^*)$. By Lemma 3,

$$y_{i*}(s, n, e^*) = y_{i*}(s', n, e^*) \leq s$$

for all $s' \geq s$. By Lemma 4, there is an $s'' \geq s$ such that

$$m_{i*}(s'', e^*) \geq m_{i*}(s, e^*) \quad \text{and}$$
$$c(s'', n) = \lim_s c(s, n) = c(n) \quad .$$

But then $c(s'', n) = P_{i*}(s'', n, e^*) = U\big(y_{i*}(s'', n, e^*)\big)$, since $n < m_{i*}(s'', e^*)$ and $y_{i*}(s'', n, e^*) \leq s \leq s''$. It follows that

$$c(n) = U\big(y_{i*}(s'', n, e^*)\big) = P_{i*}(s, n, e^*) \quad ,$$

since $y_{i*}(s'', n, e^*) = y_{i*}(s, n, e^*) \leq s$.

Let $t(n)$ denote the partial function

$$\mu s[s \geq s^* \ \&\ n < m_{i*}(s, e^*)] \quad ;$$

by Lemma 4, $t(n)$ is defined for all n. By Lemma 5,

$$c(n) = P_{i*}\big(t(n), n, e*\big)$$

for all n. Since each of the functions $m_{i*}(s, e)$ and $P_{i*}(s, n, e)$ is recursive in B, it must be that C is recursive in B. That completes our argument that for each $i < 2$ and each e, the set $\{K_i(s, e) \mid s \geq 0\}$ is finite. This is all we need to know to show C is not recursive in B, D_i for any $i < 2$. Recall that B contains only even numbers and C only odd. This means that B and D_i are each recursive in $B \cup D_i$, since

$$n \in B \longleftrightarrow (n \text{ is even } \& \ n \in B \cup D_i) \quad ,$$
$$n \in D_i \longleftrightarrow (n \text{ is odd } \& \ n \in B \cup D_i) \quad .$$

Thus it is sufficient to show C is not recursive in $B \cup D_i$ for any $i < 2$. The representing function of $B \cup D_i$ is $b \cdot d_i$.

Fix i and e. Let m be the largest member of $\{K_i(s, e) \mid s \geq 0\}$. We now show $\{e\}^{b \cdot d_i}(n)$ is either undefined or unequal to $c(n)$ for some $n \leq m$. Suppose this last is not the case. Then

$$y(n) \simeq \mu y T_1^1\Big(\prod_{j<y} p_j^{b(j) \cdot d_i(j)}, e, n, y\Big)$$

is defined and $U(y(n)) = c(n)$ for all $n \leq m$. Let y be the largest member of $\{y(n) \mid n \leq m\}$. Let s be such that $y \leq s$,

$$c(s, n) = c(n) \quad \text{and} \quad d_i(s, j) = d_i(j)$$

for all $n \leq m$ and all $j < y$. Then $y_i(s, n, e) = y(n) \leq s$ for all $n \leq m$. It follows that

$$P_i(s, n, e) = U\big(y_i(s, n, e)\big) = U\big(y(n)\big) = c(n) = c(s, n)$$

for all $n \leq m$. This means $m < m_i(s, e)$. But then by remark (R2), $y(m) \leq K_i(s, e) \leq m$. This last is absurd, since $m < y(m)$ by GND.

It remains only to see that D_i is recursively enumerable in B. It is clear that $D_i \neq 0$, since otherwise D_{1-i} would equal D and C would be recursive in B, D_{1-i}. For each $i < 2$, we define a function g_i which is recursive in B and which enumerates D_i:

$$g_i(n) = \begin{cases} (n)_1 & \text{if } d_i\big((n)_0, (n)_1\big) = 0 \\ \mu t(t \in D_i) & \text{otherwise} \end{cases} \quad .$$

We need the following elementary lemma to obtain some corollaries to Theorem 1.

LEMMA 6. If A_0 and A_1 are disjoint sets and if for each $i < 2$, A_i is recursively enumerable in $A_0 \cup A_1$, then the degree of $A_0 \cup A_1$ equals the least upper bound of the degrees of A_0 and A_1.

PROOF. Let $\underline{a}_0$, $\underline{a}_1$ and $\underline{a}_2$ be the degrees of A_0, A_1 and $A_0 \cup A_1$ respectively. Clearly $\underline{a} \leq \underline{a}_0 \cup \underline{a}_1$. We show $\underline{a}_0 \leq \underline{a}$. Suppose we wish to know if $n \in A_0$. We first ask if $n \in A_0 \cup A_1$. If the answer is no, then $n \notin A_0$. If the answer is yes, then we simultaneously enumerate A_0 and A_1; eventually n will turn up in either A_0 or A_1, but not in both, and it will then be clear whether or not $n \in A_0$. Each of the enumeration functions is recursive in $A_0 \cup A_1$. Thus $\underline{a}_0 \leq \underline{a}$, and similarly, $\underline{a}_1 \leq \underline{a}$.

COROLLARY 1. Let $\underline{b}$ and $\underline{c}$ be degrees such that $\underline{b} < \underline{c}$ and $\underline{c}$ is recursively enumerable in $\underline{b}$. Then there exist degrees $\underline{c}_0$ and $\underline{c}_1$ such that $\underline{c}_0 \cup \underline{c}_1 = \underline{c}$, $\underline{c}_0 | \underline{c}_1$ and such that for each $i < 2$, $\underline{b} < \underline{c}_i < \underline{c}$ and $\underline{c}_i$ is recursively enumerable in $\underline{b}$.

PROOF. Let B and C be sets of degrees $\underline{b}$ and $\underline{c}$ respectively, and let C be recursively enumerable in B. We set $D = C$ and apply Theorem 1. Thus we obtain disjoint sets D_0 and D_1 such that for each $i < 2$, D_i is recursively enumerable in B, and C is not recursive in B, D_i; in addition, $D = D_0 \cup D_1$. For each $i < 2$, let $\underline{d}_i$ be the degree of D_i and let $\underline{c}_i = \underline{b} \cup \underline{d}_i$. For each $i < 2$, D_i is recursively enumerable in C, since B is recursive in C. It follows from Lemma 6 that $\underline{c} = \underline{d}_0 \cup \underline{d}_1$. Then

$$\underline{c}_0 \cup \underline{c}_1 = \underline{b} \cup \underline{d}_0 \cup \underline{d}_1 = \underline{b} \cup \underline{c} = \underline{c} .$$

Since C is not recursive in B, D_i for any $i < 2$, it must be that $\underline{c}_i = \underline{b} \cup \underline{d}_i < \underline{c}$ for all $i < 2$. If $\underline{c}_0$ and $\underline{c}_1$ were comparable, then one of them would equal $\underline{c}$. Since $\underline{c}_0 | \underline{c}_1$, we have $\underline{b} < \underline{c}_i$ for all $i < 2$.

In [13] Muchnik asserted without proof that there is no minimal, non-recursive, recursively enumerable degree. In [2] Friedberg outlined a proof that if $\underline{c}$ is a non-recursive, recursively enumerable degree, then there exist incomparable, recursively enumerable degrees $\underline{c}_0$ and $\underline{c}_1$, each one of which is less than $\underline{c}$. A trivial modification of Friedberg's argument is all that is needed to prove: if $\underline{b}$ and $\underline{c}$ are degrees such that $\underline{b} < \underline{c}$ and $\underline{c}$ is recursively enumerable in $\underline{b}$, then there exist degrees $\underline{c}_0$ and $\underline{c}_1$ such that $\underline{c}_0 | \underline{c}_1$ and such that for each $i < 2$, $\underline{b} < \underline{c}_i < \underline{c}$ and $\underline{c}_i$ is recursively enumerable in $\underline{b}$. We do not know of any further modification of Friedberg's argument that will prove our Corollary 1 above.

COROLLARY 2. Each non-recursive, recursively enumerable set is the union of two disjoint, recursively enumerable sets whose degrees are incomparable.

PROOF. Let $B = 0$ and $D = C$, where C is a non-recursive, recursively enumerable set, and then apply Theorem 1. Thus $C = D_0 \cup D_1$, $D_0 \cap D_1 = 0$, and for each $i < 2$, C is not recursive in D_i and D_i is recursively enumerable. Then by Lemma 6, $\underline{c} = \underline{d}_0 \cup \underline{d}_1$, where $\underline{c}$, $\underline{d}_0$ and $\underline{d}_1$ are the degrees of C, D_0 and D_1 respectively. If $\underline{d}_0$ and $\underline{d}_1$ were comparable, then C would be recursive in D_i for some $i < 2$.

Corollary 2 improves Theorem 1 of Friedberg [3]. His theorem states that each non-recursive, recursively enumerable set is the union of two disjoint, non-recursive, recursively enumerable sets.

COROLLARY 3. If $\underline{c}$ is a degree which is recursively enumerable in some degree less than itself, then $\underline{c}$ is the least upper bound of the set of all degrees less than itself.

PROOF. Let $\underline{c}$ be recursively enumerable in $\underline{b}$, where $\underline{b} < \underline{c}$, and apply Corollary 1. Then $\underline{c} = \underline{c}_0 \cup \underline{c}_1$, and for each $i < 2$, $\underline{c}_i < \underline{c}$. If $\underline{g}$ is greater than or equal to every degree less than $\underline{c}$, then $\underline{g} \geq \underline{c}_0 \cup \underline{c}_1 = \underline{c}$.

We saw in Section 4 that there exists a recursively enumerable degree $\underline{d}$ such that $\underline{0} < \underline{d} < \underline{0}'$; clearly $\underline{d}$ is not complete, but by Corollary 3, $\underline{d}$ is the least upper bound of the set of all degrees less than $\underline{d}$. Since each complete degree is recursively enumerable in some degree less than itself, it follows that Corollary 3 strengthens a result of

Spector [25]; he showed that each complete degree is the least upper bound of the set of all degrees less than itself. The strengthening is a consequence of the fact that Spector's argument did not make use of the priority method. In Section 11 we show there is a degree $\underline{d}$ such that $\underline{d}$ is the least upper bound of the set of all degrees less than $\underline{d}$ and such that $\underline{d}$ is not recursively enumerable in some degree less than itself.

One of the questions left open by Spector in [25] was: does there exist an infinite, ascending sequence of degrees with a minimal upper bound? Corollary 4 answers this question in the affirmative. In particular, each non-recursive, recursively enumerable degree is a minimal upper bound of some infinite, ascending sequence of degrees.

COROLLARY 4. If $\underline{c}$ is a degree which is recursively enumerable in some degree less than itself, then there exists an infinite, ascending sequence of degrees which has $\underline{c}$ as a minimal upper bound.

PROOF. Let $\underline{c}$ be recursively enumerable in $\underline{b}$, where $\underline{b} < \underline{c}$. Let B be the subordering of all degrees greater than or equal to $\underline{b}$ and less than $\underline{c}$. If follows from Corollary 1 that B has no maximal member. By Hausdorff's principle, B contains a maximal, linearly ordered subset, call it T. Since B has no maximal member, T has no greatest member. Let $T = \{\underline{t}_0, \underline{t}_1, \underline{t}_2, \ldots\}$. Let $r(0) = 0$, and for each n, let $r(n+1)$ be the least k such that $\underline{t}_k > \underline{t}_{r(n)}$. Let $\underline{b}_n = \underline{t}_{r(n)}$ for all n. Then $\underline{b}_0 < \underline{b}_1 < \underline{b}_2 < \ldots < \underline{c}$. If there were a $\underline{g}$ such that $\underline{b}_0 < \underline{b}_1 < \underline{b}_2 < \ldots < \underline{g} < \underline{c}$, then $\underline{g}$ would be greater than every member of T, and T would not be maximal.

It would be interesting to know if Corollary 4 can be proved without any use of the axiom of choice.†

COROLLARY 5. If $\underline{b}$, $\underline{c}$ and $\underline{d}$ are degrees such that $\underline{b} < \underline{c} < \underline{d}$ and $\underline{d}$ is recursively enumerable in $\underline{b}$, then there exists a degree $\underline{h}$ such that $\underline{b} < \underline{h} < \underline{d}$, $\underline{c} | \underline{h}$ and $\underline{h}$ is recursively enumerable in $\underline{b}$.

---

† Note added in proof: Corollary 4 can be proved without the axiom of choice, since for each function g, the set $\{f \mid \underline{f} \leq \underline{g}\}$ has an obvious well-ordering.

PROOF. Let B, C and D be sets of degrees $\underline{b}$, $\underline{c}$ and $\underline{d}$ respectively, and let D be recursively enumerable in B. We apply Theorem 1 to obtain sets D_0 and D_1. For each $i < 2$, let $\underline{d}_i$ be the degree of D_i and let $\underline{h}_i = \underline{b} \cup \underline{d}_i$. By Lemma 6, $\underline{d} = \underline{d}_0 \cup \underline{d}_1$, and consequently $\underline{d} = \underline{h}_0 \cup \underline{h}_1$. Thus for each $i < 2$, $\underline{b} \leq \underline{h}_i \leq \underline{d}$. Since C is not recursive in B, D_i for any $i < 2$, it follows that $\underline{c} \nleq \underline{h}_i$ for any $i < 2$. If $\underline{h}_i \leq c$ for all $i < 2$, then $\underline{d} = \underline{h}_0 \cup \underline{h}_1 \leq \underline{c} < \underline{d}$; consequently, either $\underline{h}_0$ or $\underline{h}_1$ is incomparable with $\underline{c}$.

Without any use of the priority method Shoenfield [23] proved that if $\underline{b}$ and $\underline{c}$ are degrees such that $\underline{b} < \underline{c} < \underline{b}'$, then there exists a degree $\underline{h}$ such that $\underline{b} < \underline{h} < \underline{b}'$ and $\underline{c} | \underline{h}$. The next two corollaries are immediate consequences of Corollary 5.

COROLLARY 6. If $\underline{c}$ is a degree such that $\underline{0} < \underline{c} < \underline{0}'$, then there exists a recursively enumerable degree $\underline{h}$ such that $\underline{c} | \underline{h}$.

COROLLARY 7. A set is recursive if and only if it is recursive in every non-recursive, recursively enumerable set.

For each e, let ω_e denote the range of the partial recursive function

$$U(\mu y T_1(e, n, y)) \quad .$$

Then $\omega_0, \omega_1, \omega_2, \ldots$ is the standard, Kleene enumeration of all recursively enumerable sets. From Corollary 6 we know there is a function f such that for all e, if ω_e is non-recursive and has degree less than $\underline{0}'$, then the degree of $\omega_{f(e)}$ is incomparable with the degree of ω_e. We do not know if f can be chosen to be recursive.† But we can show that there exist recursive functions f and g such that for all e, if ω_e is non-recursive and has degree less than $\underline{0}'$, then at least one of the sets $\omega_{f(e)}$ and $\omega_{g(e)}$ has degree incomparable with the degree of ω_e. The existence of f and g is clear from an examination of the proof of Theorem 1. Let $B = 0$, let $C = \omega_e$, where ω_e is non-recursive, and let D be a recursively enumerable set of degree $\underline{0}'$. Theorem 1 tells us how to recursively enumerate sets D_0 and D_1 such that ω_e is not recursive

† Note added to second edition: Yates [39] has shown that f can be chosen to be recursive.

in D_i for any $i < 2$ and such that $D_0 \cup D_1 = D$. It follows, as we saw in Corollary 5, that if ω_e has degree less than $\underline{0}'$, then at least one of the sets D_0 and D_1 has degree incomparable with the degree of ω_e. The proof of Theorem 1 is such that we can effectively find recursive procedures for enumerating D_0 and D_1 if we are given a recursive procedure for enumerating ω_e.

We now generalize the concluding result of Section 4.

THEOREM 2. Let $\underline{b}$ and $\underline{d}$ be degrees such that $\underline{b} < \underline{d}$ and $\underline{d}$ is recursively enumerable in $\underline{b}$. Let T be the subordering of all degrees less than or equal to $\underline{d}$, recursively enumerable in $\underline{b}$ and greater than or equal to $\underline{b}$. Then every countable, partially ordered set is imbeddable in T.

PROOF. We combine the argument of Theorem 1 of the present section with the argument of Corollary 3 of Section 4. In what follows, we assume complete familiarity with these two arguments.

Let B and D be sets of degrees $\underline{b}$ and $\underline{d}$ respectively such that D is recursively enumerable in B. Let f be a function recursive in B which enumerates D without repetitions. Let b and d be the representing functions of B and D respectively. Let $d(s, n) = 0$ if there is a k less than s such that $f(k) = n$; otherwise let $d(s, n) = 1$. For each n, $\lim_s d(s, n)$ exists and is equal to $d(n)$. The function $d(s, n)$ is recursive in B.

Our objective is to define a sequence $D_0, D_1, D_2, \ldots$ of sets such that

(1) $\bigcup_i D_i = D$;

(2) $i \neq j \rightarrow D_i \cap D_j = 0$;

(3) D is not recursive in $B, \bigcup_{j \neq i} D_j$.

In addition we want the D_i's to be simultaneously recursively enumerable in B. This last means there must be a function $g(i, n)$ such that $g(i, n)$ is recursive in B and such that for each i,

$$D_i = \{g(i, n) \mid n \geq 0\} .$$

Our construction is quite similar to that of Theorem 1. We define six functions, $t(s)$, $d(i, s, n)$, $y(i, s, n, e)$, $P(i, s, n, e)$, $m(i, s, e)$ and $K(i, s, e)$, simultaneously by induction on s. At stage s we put $f(s)$ into just one of the D_i's thereby guaranteeing that the D_i's will be disjoint and that their union will be D. For each i, $\lim_s d(i, s, n)$ will be the representing function of D_i. Each of our six functions will be recursive in B. We assume without any loss of generality that every member of B is even and that every member of D is odd.

Stage $s = 0$. We set $d(0, 0, n) = 1$ for all $n \neq f(0)$, and we set $d(0, 0, f(0)) = 0$. We set $P(i, 0, n, e) = 2$, $d(i+1, 0, n) = t(0) = y(i, 0, n, e) = 1$ and $m(i, 0, e) = K(i, 0, e) = 0$ for all i, e and n.

Stage $s > 0$. We define $t(s)$ and $z(s)$:

$$t(s) = \mu x_{x \leq f(s)}\Big[(Ei)(Ee)\big(x = p_i^{1+e} \;\&\; f(s) < K(i, s-1, e)\big)\Big] \;;$$

$$z(s) = \mu i\Big[i < t(s) \;\&\; \big(t(s)\big)_i \neq 0\Big] \;.$$

Note that $z(s) \leq f(s)$. We set $d(z(s), s, f(s)) = 0$ and

$$d(i, s, n) = d(i, s-1, n)$$

for all i and n such that $i \neq z(s)$ or $n \neq f(s)$. Thus we have put $f(s)$ into $D_{z(s)}$. Let

$$d^i(s, j) = \prod_{k \neq i \,\&\, k \leq s} d(k, s, j) \;.$$

We define $y(i, s, n, e)$ and $P(i, s, n, e)$ for all i, n and e:

$$y(i, s, n, e) = \mu y_{y \leq s}\Big[T_1^1\Big(\prod_{j < y} p_j^{b(j) \cdot d^i(s,j)}, e, n, y\Big)\Big]$$

$$P(i, s, n, e) = \begin{cases} U\big(y(i, s, n, e)\big) & \text{if } y(i, s, n, e) \leq s \\ s+2 & \text{otherwise.} \end{cases}$$

Before we define $m(i, s, e)$, we observe that

$$(i)(e)(Et)[t \leq s \;\&\; d(s, t) \neq P(i, s, t, e)] \;.$$

This last is clear, since $d(s, s) = 1$, and since by GND, $P(i, s, s, e) = s + 2$ for all i and e:

$$m(i, s, e) = \mu t[d(s, t) \neq P(i, s, t, e)] \;;$$

$$K(i, s, e) = \begin{cases} K(i, s-1, e) & \text{if } (n)[n < m(i, s, e) \rightarrow y(i, s, n, e) \leq K(i, s-1, e)] \\ \mu t(n)[n < m(i, s, e) \rightarrow y(i, s, n, e) \leq t] & \text{otherwise .} \end{cases}$$

For each i, let $d_i(n) = \lim_s d(i, s, n)$, and let D_i be the set whose representing function is d_i; in addition let d^i be the representing function of

$$D^i = \bigcup_{j \neq i} D_j .$$

Then for each i and n, $\lim_s d^i(s, n) = d^i(n)$.

We now proceed exactly as in Theorem 1. We first show that for each i and e, the set $\{K(i, s, e) | s \geq 0\}$ is finite, and then use this fact to show D is not recursive in B, D^i for any i. Suppose there is an i and an e such that the set $\{K(i, s, e) | s \geq 0\}$ is infinite. Let

$$t^* = \mu t(Ei)(Ee)\left[t = p_i^{1+e} \ \& \ \{K(i, s, e) | s \geq 0\} \text{ is infinite}\right] .$$

Let i^* and e^* be such that

$$t^* = p_{i^*}^{1+e^*} .$$

All that is needed now is to repeat the arguments of Lemmas 1-5 in order to show D is recursive in B, which is impossible, since $\underline{b} < \underline{d}$. There would be little profit in actually repeating these arguments, since the only real difference between the constructions of Theorems 1 and 2 resides in the assignment of priorities; all other differences are merely notational. We content ourselves with proving the counterpart of Lemma 1 and stating the counterparts of Lemmas 4 and 5.

LEMMA 7. There is an s' such that for all $s \geq s'$, $z(s) = i^*$ or $t^* < t(s)$.

PROOF. Suppose there are infinitely many s such that $z(s) \neq i^*$ and $t^* \geq t(s)$. Let S be an infinite set such that for all $s \in S$, $z(s) \neq i^*$ and $f(s) \geq t^* \geq t(s)$. (Recall that f enumerates D without repetitions.) Since S is infinite, there must be an infinite subset R of S and a t^{**} such that $t(s) = t^{**}$ for all $s \in R$.

Thus we have $f(s) \geq t^* \geq t(s) = t^{**}$ for all $s \in R$. Since $f(s) \geq t(s)$ for all $s \in R$, it follows from the definition of $t(s)$ that there exist i^{**} and e^{**} such that

$$t(s) = t^{**} = p_{i^{**}}^{1+e^{**}} \quad \text{and} \quad f(s) < K(i^{**}, s-1, e^{**})$$

for all $s \in R$. But then the set $\{K(i^{**}, s, e^{**}) | s \geq 0\}$ is infinite, since R is infinite, and since f enumerates D without repetitions. This means that $t^* \leq t^{**}$, and consequently $t^* = t^{**}$. It follows that $t(s) = t^*$ for all $s \in R$. But then $z(s) = i^{**} = i^*$ for all $s \in R$, which is impossible, since $z(s) \neq i^*$ for all $s \in S$.

Let s^* be the least s' such that for all $s > s'$, $t^* \leq f(s)$ and either $z(s) = i^*$ or $t^* < t(s)$. The natural number s^* plays the same role it did in the proof of Theorem 1. There is now no difficulty in mimicing the arguments of Lemmas 2-5. Lemma 4 becomes: the set $\{m(i^*, s, e^*) | s \geq 0\}$ is infinite. Lemma 5 becomes: if $s \geq s^*$ and $n < m(i^*, s, e^*)$, then $d(n) = P(i^*, s, n, e^*)$. It then follows, as in Theorem 1, that D is recursive in B, since each of the functions $m(i^*, s, e^*)$ and $P(i^*, s, n, e^*)$ is recursive in B. Since we are given that D is not recursive in B, that completes our argument by reductio ad absurdum that for each i and e, the set $\{K(i, s, e) | s \geq 0\}$ is finite. The concluding argument of Theorem 1 is easily repeated in order to show D is not recursive in B, D^i for any i.

It remains only to see that the D_i's are simultaneously recursively enumerable in B. For each i, $D_i \neq 0$, since otherwise D^i would equal D, and D would be recursive in B, D^i. The function $g(i, n)$ is recursive in B and enumerates the D_i's simultaneously:

$$g(i, n) = \begin{cases} (n)_1 & \text{if } d(i, (n)_0, (n)_1) = 0 \\ \left(\mu t\left(d(i, (t)_0, (t)_1) = 0\right)\right)_1 & \text{otherwise} \end{cases} ;$$

$$D_i = \{g(i, n) | n \geq 0\} .$$

To see that each D_j is recursive in D, we repeat the argument of Lemma 6. Suppose we wish to know if $n \in D_j$. We first ask if $n \in D$. If the answer is no, then $n \notin D_j$. If the answer is yes, then we simultaneously enumerate the D_i's, and eventually n will turn up in exactly one of the D_i's, thereby answering our initial question. Actually, we have just shown that the D_i's are uniformly recursive in D; that is, there is a recursive function h such that D_i is recursive in D with Gödel number $h(i)$. By a similar argument, it follows that for each i, the members of the sequence

$$D_0, D_1, \ldots, D_{i-1}, D_{i+1}, \ldots$$

are uniformly recursive in D^i.

For each i, let $F_i = B \cup D_i$. Then B is recursive in F_i, because every member of B is even and every member of D_i is odd. In addition the F_i's are uniformly recursive in D and are simultaneously recursively enumerable in B. Let $\lambda mn|F(m, n)$ be a function such that for each m, $\lambda n|F(m, n)$ is the representing function of F_m. Recall that for all $s > 0$, $z(s) \leq f(s)$. This means

$$(n)(n \in D \leftrightarrow n \in \bigcup \{D_i | i \leq n\}) \quad ,$$

since for all $s > 0$, $f(s) \in D_{z(s)}$, and since $f(0) \in D_0$. It follows that D is recursive in $\lambda mn|F(m, n)$, since

$$(n)(n \in D \leftrightarrow n \text{ is odd} \,\&\, (Ei)(i \leq n \,\&\, F(i, n) = 0)) \quad .$$

We show that the F_i's are recursively independent (in sequence). For each i, let F^i denote the function

$$\lambda mn|F(m + sg((m+1) \dot{-} i), n) \ .$$

First we observe that

$$(m)(n)\Big\{F\Big(m + sg\big((m+1) \dot{-} i\big), n\Big) = 0 \leftrightarrow n \in B \vee \Big[n \in D_{m+sg((m+1)\dot{-}i)} \,\&\, n \geq m + sg\big((m+1) \dot{-} i\big)\Big]\Big\} \ .$$

It follows that F^i is recursive in B, D^i, since the sets

$$D_0, D_1, \ldots, D_{i-1}, D_{i+1}, \ldots$$

are uniformly recursive in D^i. But then D is not recursive in F^i, since D is not recursive in B, D^i. This means $\lambda mn|F(m, n)$ is not recursive in F^i, since D is recursive in $\lambda mn|F(m, n)$. It follows that F_i is not recursive in F^i.

Let $\leq_R$ be the recursive, universal, partial ordering relation described in the proof of Corollary 3 of Section 4. For each i, let

$$B_i = \left\{p_i^{1+n} | n \in F_i\right\}$$

Then the B_i's are uniformly recursive in D, simultaneously recursively enumerable in B and recursively independent (in sequence); also B is recursive in each B_i. For each u, let

$$C_u = \cup \{B_i \mid i \leq_R u\} \quad .$$

Then for each u, B is recursive in C_u, C_u is recursive in D and C_u is recursively enumerable in B.

We conclude the proof of our theorem in exactly the same manner that we concluded the proof of Corollary 3 of Section 4. It is necessary to show

$$(u)(v)(u \leq_R v \longleftrightarrow C_u \text{ is recursive in } C_v) \quad .$$

If $u \not\leq_R v$, then C_u is not recursive in C_v because of the recursive independence of the B_i's. If $u \leq_R v$, then C_u is recursive in C_v because of the recursiveness of $\leq_R$.

§6. THE JUMP OPERATOR

The jump operator for degrees was defined in Section 1; we recall some of its properties:

(i) $\underline{d} < \underline{d}'$;

(ii) $\underline{b} \leq \underline{d} \rightarrow \underline{b}' \leq \underline{d}'$;

(iii) if $\underline{h}$ is recursively enumerable in $\underline{d}$, then $\underline{h} \leq \underline{d}'$;

(iv) $\underline{d}'$ is recursively enumerable in $\underline{d}$.

Our main purpose in the present section is to study the effect of the jump operator on the ordering of degrees. The trend of our results may be put intuitively as follows: the jump operator has almost no respect for the ordering of degrees. For example, we will see that the converse of (ii) is false, and remains false even when fairly stringent conditions are placed on $\underline{b}$ and $\underline{d}$.

Theorem 1 is adapted from Friedberg [4]. Our version of Friedberg's argument is motivated by the hope that it will make the proofs of Theorems 2 and 3 clearer. Recall that a degree $\underline{c}$ is said to be complete if there is a degree $\underline{d}$ such that $\underline{d}' = \underline{c}$.

THEOREM 1. A degree $\underline{c}$ is complete if and only if $\underline{c} \geq \underline{0}'$.

PROOF. We prove the "if" portion of the theorem. Let c be the representing function of a set of degree $\underline{c}$. We intend to define a function d such that the degree of the predicate

$$(Ey)T_1^1\left(\tilde{d}(y),\ e,\ e,\ y\right)$$

is $\underline{c}$. First we introduce a recursive predicate K that will be needed in the definition of d:

$$K(u, v, e, d, h, c) \leftrightarrow \Big\{T_1^1(u, e, e, v) \ \& \ v \le \ell h(u)$$
$$\& \ (i)_{i < h}\big((u)_i = (d)_i\big)$$
$$\& \ (i)(t)_{t < e}(m)_{m > 0}\Big[h \le i < \ell h(u) \ \& \ p_t^m = i \rightarrow (u)_i = (c)_t\Big]\Big\}$$

It is clear that the predicate $(Ey)K\big((y)_0, (y)_1, e, d, h, c\big)$ has degree less than or equal to $\underline{0}'$. It follows the predicate

$$(Ey)K\big((y)_0, (y)_1, e, d, h, \tilde{c}(e)\big)$$

is recursive in the function c, since $\underline{c} \ge \underline{0}'$.

Our plan is to define three functions, $y(e)$, $h(e)$, and $d(e, i)$, simultaneously by induction on e; each of these functions will be recursive in c. In addition, the function $d(e, i)$ will have the property that

$$d(i, i) = d(e, i)$$

whenever $e \ge i$. The desired function $d(i)$ will be defined by

$$d(i) = d(i, i) \quad .$$

Stage $e = 0$. We set $y(0) = h(0) = 1$. We define $d(0, i)$ for all i:

$$d(0, i) = \begin{cases} c(0) & \text{if } (Em)(p_0^m = i \ \& \ m > 0) \\ 1 & \text{otherwise} \quad . \end{cases}$$

Stage $e > 0$. We set:

$$y(e) = \begin{cases} \mu y K\big((y)_0, (y)_1, e, \big(\tilde{d}(e, h(e-1))\big)_{e-1}, h(e-1), \tilde{c}(e)\big) \\ \qquad \text{if such a } y \text{ exists,} \\ 1 \ \text{otherwise} \ ; \ h(e) = h(e-1) + y(e). \end{cases}$$

Note that $\ell h\big((y(e))_0\big) \le h(e)$. With this last in mind, we define $d(e, i)$ for all i:

$$d(e, i) = \begin{cases} \big(y(e)\big)_{0,i} & \text{if } h(e-1) \le i < \ell h\big((y(e))_0\big) \\ c(e) & \text{if } h(e) \le i \ \& \ (Em)_{m > 0}(i = p_e^m) \\ d(e-1, i) & \text{otherwise} \quad . \end{cases}$$

That concludes the construction. Since $y(e) > 0$ for all e, we have $h(e) > e$ for all e. Since $h(e-1) < h(e)$ whenever $e > 0$, it follows

$d(e, i) = d(e-1, i)$ whenever $i \leq e-1$. But then

$$d(e, i) = d(i, i)$$

whenever $e \geq i$. We define $d(i) = d(i, i)$. Let $\underline{d}$ be the degree of $d(i)$.

LEMMA 1. $\underline{c} \leq \underline{d}'$.

PROOF. It is clear from the construction that

$$d(t, p_t^m) = c(t)$$

whenever $h(t) \leq p_t^m$ and $m > 0$. We show

$$d(e, p_t^m) = c(t)$$

whenever $h(t) \leq p_t^m$, $m > 0$ and $t \leq e$ by an induction on e. Let t, e and m be such that $h(t) \leq p_t^m$, $m > 0$ and $e > t$, and suppose

$$d(e-1, p_t^m) = c(t)$$

Let $p_t^m = i$. Now $d(e, i)$ equals either $d(e-1, i)$ or $\big(y(e)\big)_{0,i}$, since $i \neq p_e^m$. Suppose $d(e, i) \neq d(e-1, i)$. Then we have

$$h(e-1) \leq i < \ell h\Big(\big(y(e)\big)_0\Big) \;\&\; p_t^m = i \;\&\; t < e \;\&\; y(e) \neq 1 \quad ,$$

since the first case of the definition of $d(e, i)$ must hold. But then it follows from the definition of K and $y(e)$ that

$$\big(y(e)\big)_{0,i} = c(t) \quad ,$$

since $y(e) \neq 1$.

Thus $d(e, p_t^m) = c(t)$ whenever $h(t) \leq p_t^m$, $m > 0$ and $t \leq e$. It follows

$$d(p_t^m) = d(p_t^m, p_t^m) = c(t)$$

whenever $h(t) \leq p_t^m$ and $m > 0$, since $p_t > t$ for all t. Then we have

$$(t)(En)\Big[(m)_{m \geq n}\big(d(p_t^m) = 0\big) \vee (m)_{m \geq n}\big(d(p_t^m) = 1\big)\Big] \quad ,$$

since $c(t) \leq 1$ for all t; for each t, let $k(t)$ be the least such n. It is clear that k has degree less than or equal to $\underline{d}'$, and that

$$c(t) = d\big(p_t^{k(t)}\big)$$

LEMMA 2. $\underline{d}' \leq \underline{c}$.

PROOF. Since y is recursive in c, it is sufficient to show

$$(Ey)T_1^1\left(\tilde{d}(y),\ e,\ e,\ y\right) \longleftrightarrow y(e) \neq 1$$

for all $e > 0$.

Fix $e > 0$ and suppose $y(e) \neq 1$. We show $\left(y(e)\right)_{0,i} = d(i)$ for all $i < \ell h\left((y(e))_0\right)$. Fix $i < \ell h\left((y(e))_0\right)$. Suppose $t > e$ and $d(t-1, i) = d(e, i)$. Then

$$i < \ell h\left((y(e))_0\right) \leq y(e) \leq h(e) \leq h(t-1) \quad ,$$

since $t > e$ and h is non-decreasing. But then it follows from the definition of $d(t, i)$ that $d(t, i) = d(t-1, i)$. Thus we have shown by induction on t that $d(t, i) = d(e, i)$ for all $t > e$. It follows $d(i) = d(e, i)$. If $h(e-1) \leq i$, then $d(e, i) = \left(y(e)\right)_{0,i}$. Suppose $i < h(e-1)$. Since $y(e) \neq 1$, we must have $\left(y(e)\right)_{0,i} = d(e-1, i)$. Since $i < h(e-1) \leq h(e)$, we must have $d(e, i) = d(e-1, i)$. Thus $d(i) = \left(y(e)\right)_{0,i}$.

Since $y(e) \neq 1$, it follows

$$T_1^1\left((y(e))_0,\ e,\ e,\ (y(e))_1\right) \ \&\ \left(y(e)\right)_1 \leq \ell h\left((y(e))_0\right) \quad .$$

We know $d(i) = \left(y(e)\right)_{0,i}$ for all $i < \ell h\left((y(e))_0\right)$. It follows

$$(Ey)T_1^1\left(\tilde{d}(y),\ e,\ e,\ y\right) \quad .$$

That completes the first half of the proof of Lemma 2. Now suppose y has the property that $T_1^1\left(\tilde{d}(y),\ e,\ e,\ y\right)$. Again let $e > 0$. We wish to show $y(e) \neq 1$. Since 0 is not the Gödel number of a deduction, it will be sufficient to show

$$K\left(\tilde{d}(y),\ y,\ e,\ \left(\tilde{d}(e,\ h(e-1))\right)_{e-1},\ h(e-1), \tilde{c}(e)\right) \quad .$$

But then it will be enough to prove:

(1) $(i)\left[i < h(e-1) \rightarrow d(i) = d(e-1, i)\right]$;

(2) $(i)(t)_{t < e}(m)_{m > 0}\left[h(e-1) \leq i < y \ \&\ p_t^m = i \rightarrow d(i) = c(t)\right]$.

To prove (1), suppose $i < h(e-1)$. Then $i < h(t-1)$ for all $t \geq e$, and consequently, $d(t, i) = d(t-1, i)$ for all $t \geq e$. It follows $d(i) = d(e-1, i)$. To prove (2), suppose $h(e-1) \leq i < y$, $t < e$, $p_t^m = i$ and

$m > 0$. Then $h(t) \leq p_t^m$, since $t < e$. It was shown in the proof of Lemma 1 that $d(p_t^m) = c(t)$ whenever $h(t) \leq p_t^m$ and $m > 0$.

By Theorem 1, the equation $\underline{x}' = \underline{c}$ has a solution if and only if $\underline{c}' \geq \underline{0}'$. It is now natural to ask: does $\underline{x}' = \underline{c}$ have a unique solution? Spector [25] showed that the answer is no regardless of the value of $\underline{c}$; his result states: for each degree $\underline{b}$, there are degrees $\underline{b}_0$ and $\underline{b}_1$ such that $\underline{b} \leq \underline{b}_0$, $\underline{b} \leq \underline{b}_1$ and $\underline{b}_0 \cup \underline{b}_1 = \underline{b}' = \underline{b}_0' = \underline{b}_1'$. We combine the idea of his proof with the system of priorities of Theorem 1 of Section 5 in order to prove Theorem 2 of the present section. We could, if we wished, obtain Theorem 2 as a complicated corollary of Theorem 1 of Section 4; however, the price of such elegance would be a total lack of clarity.

> THEOREM 2. Let A and B be sets such that B is recursively enumerable in A. Then there exist disjoint sets B_0 and B_1 such that $B_0 \cup B_1 = B$ and such that for each $i < 2$, $(B_i \cup A)'$ is recursive in A', and B_i is recursively enumerable in A.

PROOF. Let f be a one-one function recursive in A whose range is $B - A$. We will define six functions simultaneously by induction on s: $b_i(s, n)$, $y_i(s, n)$ and $t_i(s)$ $(i = 0, 1)$. Each of these functions will be recursive in A. For each $i < 2$, we will have

$$1 \geq b_i(s, n) \geq b_i(s + 1, n) \geq 0$$

for all s and n. It follows from this last (cf. Section 5) that for each $i < 2$, $\lim_s b_i(s, n)$ is the representing function of a set recursively enumerable in A, namely, C_i. At stage s of the construction, we put $f(s)$ (the sth member of $B - A$) in C_0 or C_1 but not in both. We assume $B - A$ is infinite, since otherwise there is nothing to prove.

Stage $s = 0$. We set $y_i(0, n) = t_i(0) = 0$ for all $i < 2$ and all n. We set $b_i(0, n) = 1$ for all $i < 2$ and all $n \notin A \cup \{f(0)\}$. We set $b_0(0, f(0)) = 1$ and $b_1(0, f(0)) = 0$. We set $b_i(0, n) = 0$ for all $i < 2$ and all $n \in A$.

Stage $s > 0$. For each $i < 2$, we define

$$t_i(s) = \mu n_{n < s}[f(s) < y_i(s-1, n)] \quad .$$

Let $z(s) = 1$ if $t_0(s) \leq t_1(s)$, and let $z(s) = 0$ otherwise. We define $b_i(s, n)$ for all $i < 2$ and all n:

$$b_i(s, n) = \begin{cases} 0 & \text{if } i = z(s) \ \& \ n = f(s) \\ b_i(s - 1, n) & \text{otherwise.} \end{cases}$$

We conclude the construction by defining $y_i(s, n)$ for all $i < 2$ and all n:

$$y_i(s, n) = \begin{cases} \mu y T_1^1\Big(\prod_{j<y} p_j^{b_i(s,j)}, n, n, y\Big) \\ \quad \text{if } (Ey)_{y \leq s}\, T_1^1\Big(\prod_{j<y} p_j^{b_i(s,j)}, n, n, y\Big) \\ 0 \ \text{otherwise} . \end{cases}$$

Since all applications of the least number operator at stage $s > 0$ are bounded, it follows that each of the functions defined above is recursive in f, hence in A. For each $i < 2$, let

$$b_i(n) = \lim_s b_i(s, n)$$

for all n, and let C_i be the set whose representing function is b_i. Let $B_0 = C_0 - A$ and $B_1 = C_1 - (A - B)$. Then $C_i = B_i \cup A$ for all $i < 2$. Since f is one-one and enumerates $B - A$, we have $B_0 \cap B_1 = 0$ and $B_0 \cup B_1 = B$. We need Lemma 3 to show $(B_i \cup A)' = C_i'$ is recursive in A'.

LEMMA 3. For each $i < 2$ and each n, $\lim_s y_i(s, n)$ exists.

PROOF. We use the method of infinite descent. Let i and n be such that $i < 2$ and $\lim_s y_i(s, n)$ does not exist. We will find a k and an m such that $k < 2$, $\lim_s y_k(s, m)$ does not exist and $2m + k < 2n + i$. There must be an infinite set S such that for each $s \in S$,

$$s > n + 1 \ \& \ y_i(s, n) \neq y_i(s - 1, n) > 0 \quad ,$$

since $\lim_s y_i(s, n)$ does not exist. We claim $t_{1-i}(s) \leq t_i(s) \leq n$ for all $s \in S$. Fix $s \in S$. Then

$$y_i(s - 1, n) = \mu y T_1^1\Big(\prod_{j<y} p_j^{b_i(s-1,j)}, n, n, y\Big) \quad ,$$

since $s - 1 > 0$ and $y_i(s - 1, n) > 0$. There must be a $j < y_i(s - 1, n)$

such that $b_i(s - 1, j) \neq b_i(s, j)$, because $y_i(s, n) \neq y_i(s - 1, n)$. But then $j = f(s)$, $f(s) < y_i(s - 1, n)$ and

$$b_i(s, f(s)) \neq b_i(s - 1, f(s)) \quad .$$

This last can happen only if $i = z(s)$. It follows $t_{1-i}(s) \leq t_i(s)$. Since $f(s) < y_i(s - 1, n)$ and $n < s$, we must have $t_i(s) \leq n$.

Since $0 \leq t_{1-i}(s) \leq n$ for all $s \in S$, there is an m and an infinite subset T of S such that

$$t_{1-i}(s) = m \leq n$$

for all $s \in T$. Let $k = 1 - i$. We show $2m + k < 2n + i$. If $m < n$, then $2m + k \leq 2m + 1 < 2n \leq 2n + i$. Suppose $m = n$. Then $t_{1-i}(s) = t_i(s) = n$ for all $s \in T$, and consequently, $z(s) = 1$ for all $s \in T$. We saw above that $z(s) = i$ for all $s \in S$. It follows $i = 1$, $k = 0$, and $2m + k < 2n + i$.

It remains only to see that $\lim_s y_k(s, m)$ does not exist. It suffices to show the set $\{y_k(s, m) \mid s \geq 0\}$ is infinite. For each $s \in T$, we have

$$t_k(s) = m \leq n < s \quad .$$

But then $f(s) < y_k(s, m)$ for all $s \in T$. Since f is one-one and T is infinite, it follows the set $\{y_k(s, m) \mid s \geq 0\}$ is infinite.

Now we fix $i < 2$ and show C_i' is recursive in A'. Lemma 3 tells us that

$$(n)(Et)(s)_{s \geq t}\bigl(y_i(s, n) = y_i(s + 1, n)\bigr) \quad .$$

We define:

$$t(n) = \mu t(s)_{s \geq t}\bigl(y_i(s, n) = y_i(s + 1, n)\bigr) \quad ;$$

$$y_i(n) = y_i\bigl(t(n), n\bigr) \quad .$$

The function $y_i(n)$ is recursive in A' because the function $y_i(s, n)$ is recursive in A. Clearly, $y_i(n) = \lim_s y_i(s, n)$ for all n. C_i' was defined in Section 1; it is the representing set of the predicate

$$(Ey)T_1^1\bigl(\widetilde{b}_i(y), n, n, y\bigr) \quad .$$

We show C_i' is recursive in A' by showing

$$(Ey)T_1^1\left(\tilde{b}_i(y),\ n,\ n,\ y\right) \longleftrightarrow y_i(n) > 0$$

for all n.

Fix n. First suppose $y_i(n) > 0$. Then

$$y_i(s + 1, n) = y(s, n) > 0$$

for all $s \geq t(n)$. Since 0 is not the Gödel number of a deduction, it follows

$$y_i(n) = y_i(s, n) = y_i\left(t(n), n\right) = \mu y T_1^1\left(\prod_{j<y} p_j^{b_i(s,j)},\ n,\ n,\ y\right)$$

for all $s \geq t(n)$. Let s' be so large that $s' \geq t(n)$ and

$$b_i(s', j) = b_i(j)$$

for all $j < y_i(n)$. Then we have

$$T_1^1\left(\tilde{b}_i(y_i(n)),\ n,\ n,\ y_i(n)\right) \quad .$$

Now suppose $(Ey)T_1^1\left(\tilde{b}_i(y),\ n,\ n,\ y\right)$. Let

$$y' = \mu y T_1^1\left(\tilde{b}_i(y),\ n,\ n,\ y\right) > 0 \quad .$$

Let t be so large that $t \geq y'$ and

$$b_i(t, j) = b_i(j)$$

for all $j < y'$. Then

$$y_i(s, n) = y_i(t, n) = y' > 0$$

for all $s \geq t$. This last means $y_i(n) = y' > 0$.

COROLLARY 1. Let $\underline{a}$ and $\underline{b}$ be degrees such that $\underline{a} < \underline{b}$ and $\underline{b}$ is recursively enumerable in $\underline{a}$. Then there exist degrees $\underline{b}_0$ and $\underline{b}_1$ such that $\underline{b}_0 \cup \underline{b}_1 = \underline{b}$, $\underline{a}' = \underline{b}_0' = \underline{b}_1'$, and such that for each $i < 2$, $\underline{a} \leq \underline{b}_i$ and $\underline{b}_i$ is recursively enumerable in $\underline{a}$.

PROOF. Let A and B be sets such that A has degree $\underline{a}$, B has degree $\underline{b}$, B is recursively enumerable in A, A has only even members and B has only odd. By Theorem 2, there are disjoint sets B_0 and B_1 such that $B_0 \cup B_1 = B$ and such that for each $i < 2$, $(B_i \cup A)'$ is recursive in A', and B_i is recursively enumerable in A. For each $i < 2$, let $\underline{b}_i$ be the degree of $B_i \cup A$. Then for each $i < 2$, $\underline{a} \leq \underline{b}_i$ and $\underline{b}_i$ is recursively enumerable in $\underline{a}$, since A and B_i are recursively separable.[†] By Lemma 6 of Section 5, the degree of B equals the union of the degrees of B_0 and B_1. It follows that $\underline{b} = \underline{b}_0 \cup \underline{b}_1$, since A and B are recursively separable and $\underline{a} \leq \underline{b}$. For each $i < 2$, we have $\underline{b}_i' \leq \underline{a}'$, because $(B_i \cup A)'$ is recursive in A'.

Corollary 1 for the special case $\underline{b} = \underline{a}'$ was first proved by Friedberg in an unpublished manuscript.

COROLLARY 2. If $\underline{a}$ and $\underline{b}$ are degrees such that $\underline{a} < \underline{b}$ and $\underline{b}$ is recursively enumerable in $\underline{a}$, then there exists a degree $\underline{c}$ recursively enumerable in $\underline{a}$ such that $\underline{a} < \underline{c} < \underline{b}$ and $\underline{c}' = \underline{a}'$.

PROOF. By Corollary 1 of Section 5, there is a degree $\underline{d}$ recursively enumerable in $\underline{a}$ such that $\underline{a} < \underline{d} < \underline{b}$. By Corollary 1 of the present section, there are degrees $\underline{d}_0$ and $\underline{d}_1$, each of which is recursively enumerable in $\underline{a}$ and greater than or equal to $\underline{a}$, such that $\underline{d}_0 \cup \underline{d}_1 = \underline{d}$ and $\underline{a}' = \underline{d}_0' = \underline{d}_1'$. Then for each $i < 2$, we have $\underline{a} \leq \underline{d}_i < \underline{b}$, since $\underline{d} < \underline{b}$. Either $\underline{a} < \underline{d}_0$ or $\underline{a} < \underline{d}_1$, since $\underline{a} < \underline{d} = \underline{d}_0 \cup \underline{d}_1$.

For each degree $\underline{b}$, let $R(\underline{b})$ be the set of all degrees greater than or equal to $\underline{b}$, recursively enumerable in $\underline{b}$, and less than or equal to $\underline{b}'$. It follows from the properties of the jump operator listed at the beginning of the present section that the image of $R(\underline{b})$ under the jump operator is a subset of $R(\underline{b}')$. We will prove as a corollary to Theorem 3 that the jump operator maps $R(\underline{b})$ onto $R(\underline{b}')$. The proof of Theorem 3 is perhaps the most complicated one occurring in this monograph; the ideas underlying the proof should be regarded as a further stage in the evolution of the ideas underlying the proof of Theorem 2. The priority method is the tool that made it possible to prove Theorem 2. Conflicting "requirements" were brought under control by means of an assignment of priorities. The proof of Lemma 3 amounted to showing that each "requirement" was "injured" only finitely often in the course of the construction. In each

---

† Two sets are recursively separable if there exists a recursive set which contains one and does not intersect the other.

manifestation of the priority method in Section 4 and 5, we showed that each "requirement" is "injured" only finitely often, and then used this fact to show each requirement either is "met" or is not "dense." In Theorem 3 we again have conflicting "requirements" to which we assign priorities, but it is no longer true that each "requirement" is "injured" finitely often. In fact, infinitely many of the "requirements" will be "injured" infinitely often. As a result we are forced to resort to a combinatorial argument more complicated than any occurring above in order to show that each "requirement" either is "met" or else is not "dense." Theorem 3 is a strengthening of the principal result of [21].

THEOREM 3. Let $\underline{a}$, $\underline{b}$, $\underline{c}$ and $\underline{g}$ be degrees such that $\underline{a} \leq \underline{b}'$, $\underline{b} \leq \underline{g}$, $\underline{g}' \leq \underline{c}$, $\underline{a} \not\leq \underline{g}$, $\underline{g}$ is recursively enumerable in $\underline{b}$, and $\underline{c}$ is recursively enumerable in $\underline{b}'$. Then there exists a degree $\underline{d}$ such that $\underline{g} \leq \underline{d}$, $\underline{d}$ is recursively enumerable in $\underline{b}$, $\underline{a} \not\leq \underline{d}$ and $\underline{d}' = \underline{c}$.

PROOF. We first prove the theorem when $\underline{b} = \underline{0}$, and then we indicate the minor alterations required when $\underline{b} > \underline{0}$. Thus we have degrees $\underline{a}$, $\underline{c}$ and $\underline{g}$ such that $\underline{a} \leq \underline{0}'$, $\underline{g}' \leq \underline{c}$, $\underline{a} \not\leq \underline{g}$, $\underline{g}$ is recursively enumerable, and $\underline{c}$ is recursively enumerable in $\underline{0}'$. We wish to find a recursively enumerable degree $\underline{d}$ such that $\underline{g} \leq \underline{d}$, $\underline{a} \not\leq \underline{d}$ and $\underline{d}' = \underline{c}$.

Let f be a function of degree $\underline{0}'$ whose range is a set C of degree $\underline{c}$. Let c be the representing function of C. Let q be a recursive function whose range is a set J of degree $\underline{0}'$. Let j be the representing function of J. We define

$$j(s, n) = \begin{cases} 0 & \text{if } (Ek)_{k<s}\big(q(k) = n\big) \\ 1 & \text{otherwise} \end{cases} .$$

It is clear that $j(s, n)$ is a recursive function, and that for each n, $\lim_s j(s, n)$ exists and is equal to $j(n)$. Since f is recursive in j, there is a Gödel number z_1 such that

$$f(n) = \{z_1\}^{j}(n) = U\Big(\mu y T_1^1\big(\tilde{j}(y),\ z_1,\ n,\ y\big)\Big)$$

for all n.

We define a recursive function, $f(s, n)$, of supreme importance to our argument.

$$f(s, n) = \begin{cases} U\left(\mu y T_1^1\left(\prod_{i<y} p_i^{j(s,i)}, z_1, n, y\right)\right) \\ \qquad \text{if } (Ey)_{y \le s} T_1^1\left(\prod_{i<y} p^{j(s,i)}, z_1, n, y\right) \\ s + 1 \quad \text{otherwise} \quad . \end{cases}$$

We claim that $\lim_s f(s, n)$ exists and is equal to $f(n)$ for all n. Our claim is a consequence of the fact that $f(n) = \{z_1\}^j(n)$ and $\lim_s j(s, n) = j(n)$ for all n. We define two recursive functions, $t(s, n)$ and $h(s, n)$, by induction on s:

$$t(s, n) = \mu m_{m<s}\left(f(s, m) = n\right) \quad ;$$

$$h(0, n) = 0 \quad ;$$

$$h(s + 1, n) = h(s, n) + sg\left(|t(s + 1, n) - t(s, n)|\right) \quad .$$

LEMMA 4. For each n, $c(n) = 0$ if and only if the set $\{h(s, n) | s \geq 0\}$ is finite.

PROOF. Fix n. First suppose $c(n) = 0$. Then there is a k such that $f(k) = n$, since f enumerates C. Let m be the least k such that $f(k) = n$. There is an s' such that $f(s, w) = f(s', w)$ whenever $s \geq s'$ and $w \leq m$, since $\lim_s f(s, w)$ exists for all w. Assume $s' > n$. Then $t(s, n) = m$ and $h(s + 1, n) = h(s', n)$ for all $s \geq s'$.

Now suppose the set $\{h(s, n) | s \geq 0\}$ is finite. Then there must be a t such that $t(s, n) \leq t$ for all s. Let u be so large that $u > t$ and $f(u, w) = f(w)$ for all w such that $w \leq t$. Then $f(u, t(u, n)) = n$, since $t(u, n) \leq t < u$. (Recall the definition of the bounded least number operator.) But

$$f(u, t(u, n)) = f(t(u, n)) \quad ,$$

because $t(u, n) \leq t$. Thus $f(t(u, n)) = n$, and consequently, $c(n) = 0$.

Let a be an everywhere positive function of degree $\underline{a}$, and let z_2 be a Gödel number such that $\{z_2\}^j(n) = a(n)$ for all n. We define

$$a(s, n) = \begin{cases} U\left(\mu y T_1^1\left(\prod_{i<y} p_i^{j(s,i)}, z_2, n, y\right)\right) \\ \qquad \text{if } (Ey)_{y \le s}\left[T_1^1\left(\prod_{i<y} p_i^{j(s,i)}, z_2, n, y\right) \,\&\, U(y) \geq 1\right] \\ 1 \quad \text{otherwise} \quad . \end{cases}$$

Note $a(s, n) \geq 1$ for all s and n. The function $a(s, n)$ is recursive; for each n, $\lim_s a(s, n)$ exists and is equal to $a(n)$. Let g be an everywhere positive, recursive function whose range is a set G of degree $\underline{g}$.

We will define four recursive functions, $y(s, n, e)$, $m(s, e)$, $r(s, n, e)$ and $d(s, n)$, simultaneously by induction on s. The function $d(s, n)$ will be such that

$$0 \leq d(s + 1, n) \leq d(s, n) \leq 1$$

for all s and n. Thus for each n, $\lim_s d(s, n)$ will exist; in addition, $\lim_s d(s, n)$ will be the representing function of a recursively enumerable set D. The degree of D will be the desired degree $\underline{d}$. At stage s of the construction, we add members to D in order to insure that $\underline{g} \leq \underline{d}$ and $\underline{c} \leq \underline{d}'$; however, with the help of a system of priorities, we exercise some restraint in order to make sure that $\underline{a} \not\leq \underline{d}$ and $\underline{d}' \leq \underline{c}$.

Stage $s = 0$. We set $y(0, n, e) = r(0, n, e) = 0$ and $m(0, e) = e + 1$ for all n and e. We set $d(0, 5 \cdot 7^{g(0)}) = 0$ and $d(0, n) = 1$ for all $n \neq 5 \cdot 7^{g(0)}$.

Stage $s > 0$. For each n and e, we define

$$y(s, n, e) = \begin{cases} \mu y T_1^1\left(\prod_{i<y} p_i^{d(s-1,i)}, e, n, y\right) \\ \qquad \text{if } n \geq e \;\&\; (Ey)_{y \leq s} T_1^1\left(\prod_{i<y} p_i^{d(s-1,i)}, e, n, y\right) \\ 0 \quad \text{otherwise} . \end{cases}$$

We define $m(s, e)$ for all e; there are three mutually exclusive cases:

CASE 1. $y(s, e, e) = 0$. We set $m(s, e) = e + 1$.

CASE 2. $y(s, e, e) > 0$ and there is an n with the property that $e < n < m(s - 1, e) \;\&\; y(s, n, e) \neq y(s - 1, n, e) \;\&\; a(s, n) \neq U(y(s, n, e))$. We set $m(s, e)$ equal to the least n with the above property.

CASE 3. Otherwise. We set

$$m(s, e) = \mu n\Big[m(s - 1, e) \leq n < 2m(s - 1, e) + s \\ \;\&\; (Et)\big(e < t \leq n \;\&\; a(s, t) \neq U\big(y(s, t, e)\big)\big)\Big] .$$

Note that if Case 3 holds and

$$(t)\Big[e < t < 2m(s-1, e) + s \rightarrow a(s, t) = U\big(y(s, t, e)\big)\Big] ,$$

then $m(s, e) = m(s-1, e) + s$; this last is a consequence of the definition of the bounded least number operator.

For each n which is neither a power of a prime nor equal to $5 \cdot 7^{g(s)}$, we set $d(s, n) = d(s-1, n)$. We set $d(s, 5 \cdot 7^{g(s)})$ equal to 0. We define $r(s, n, e)$ and $d(s, p_e^n)$ for all n and e by means of a simultaneous induction on e. Let $e \geq 0$ and suppose $r(s, n, i)$ and $d(s, p_i^n)$ have been defined for all $i < e$ and all n; we define $r(s, n, e)$ and $d(s, p_e^n)$ for all n as follows:

$$r(s, n, e) = \begin{cases} 0 & \text{if } (Ei)(Em)(Et)\Big[i < e \leq t \leq n \\ & \quad \& \ p_i^m < y(s, t, e) \\ & \quad \& \ d(s, p_i^m) \neq d(s-1, p_i^m)\Big] \\ 0 & \text{if } (Et)\Big[e \leq t \leq n \\ & \quad \& \ 5 \cdot 7^{g(s)} < y(s, t, e) \\ & \quad \& \ d(s, 5 \cdot 7^{g(s)}) \neq d(s-1, 5 \cdot 7^{g(s)})\Big] \\ 1 & \text{otherwise ;} \end{cases}$$

$$d(s, p_e^n) = \begin{cases} d(s-1, p_e^n) & \text{if } n \geq h(s, e) \\ d(s-1, p_e^n) & \text{if } (Ei)(Em)\Big[i \leq e \\ & \quad \& \ i \leq m < m(s, i) \ \& \ r(s, m, i) = 1 \\ & \quad \& \ p_e^n < y(s, m, i)\Big] \\ 0 & \text{otherwise .} \end{cases}$$

It is easily verified that each of the four functions just defined is recursive. Such a verification is possible for two reasons: each of the functions $a(s, n)$, $h(s, n)$ and $g(s)$ is recursive; all quantifiers, as well as all instances of the least operator, are properly bounded. For each n, let

$$d(n) = \lim_s d(s, n) \quad ;$$

then $d(n) = 0$ if and only if there is an s such that $d(s, n) = 0$. Thus d is the representing function of a recursively enumerable set. Let $\underline{d}$ be the degree of d. We have $\underline{g} \leq \underline{d}$, since for all $n > 0$, we have

$$d(5 \cdot 7^n) = 0 \longleftrightarrow n \in G \quad .$$

We list three remarks needed in some vital parts of the main body of our argument:

(R1) $(s)(e)(m(s, e) > e)$;

(R2) $(s)(n)(e)[r(s, n, e) = 0 \rightarrow r(s, n + 1, e) = 0]$;

(R3) $(s)(n)(e)[y(s, n, e) = 0 \mathbin{\&} n > e \rightarrow m(s, e) \leq n]$.

Remark (R1) is readily proved by induction on s if the definition of the bounded least number operator is kept in mind. Remark (R2) is an immediate consequence of the definition of $r(s, n, e)$.

We prove (R3) by induction on s. We know

$$(n)(e)[y(0, n, e) = 0 \mathbin{\&} n > e \rightarrow m(0, e) \leq n] \quad .$$

Let s be such that $s > 0$ and

$$(n)(e)[y(s - 1, n, e) = 0 \mathbin{\&} n > e \rightarrow m(s - 1, e) \leq n] \quad .$$

Let e and n be such that

$$y(s, n, e) = 0 \mathbin{\&} n > e \quad .$$

Then $a(s, n) \neq U(y(s, n, e))$, since $a(s, n) \geq 1$ and $U(0) = 0$. First we suppose $n < m(s - 1, e)$. Then $y(s - 1, n, e) > 0$ by the induction hypothesis. Thus we have

$$e < n < m(s - 1, e) \mathbin{\&} y(s, n, e) \neq y(s - 1, n, e) \mathbin{\&} a(s, n) \neq U(y(s, n, e)) \quad .$$

Either Case 1 or Case 2 of the definition of $m(s, e)$ must hold; in either event, $m(s, e) \leq n$. Now we suppose $m(s - 1, e) \leq n$. If either Case 1 or Case 2 of the definition of $m(s, e)$ holds, then we have $m(s, e) \leq m(s - 1, e) \leq n$ by remark (R1). If Case 3 holds and $n < 2m(s - 1, e) + s$, then $m(s, e) \leq n$, since $a(s, n) \neq U(y(s, n, e))$. If Case 3 holds and $n \geq 2m(s - 1, e) + s$, then $m(s, e) \leq 2m(s - 1, e) + s \leq n$.

LEMMA 5. Let $y(s, n, e) > 0$, $m(s, e) > n \geq e$ and $r(s, n, e) = 1$. Then $y(s, n, e) = y(s + 1, n, e)$.

PROOF. Since $y(s, n, e) > 0$, we have $s > 0$ and

$$y(s, n, e) = \mu y T_1^1\Big(\prod_{i<y} p_i^{d(s-1,i)}, e, n, y\Big) .$$

We suppose $y(s, n, e) \neq y(s+1, n, e)$ and then show $r(s, n, e) = 0$. Since $y(s+1, n, e) \neq y(s, n, e) > 0$, there must be a $j < y(s, n, e)$ such that $d(s, j) \neq d(s-1, j)$. If $j = 5 \cdot 7^{g(s)}$, then $r(s, n, e) = 0$. Suppose $j \neq 5 \cdot 7^{g(s)}$. Then j must be a prime power; let $j = p_i^m$. If $i < e$, then $r(s, n, e) = 0$. Suppose $i \geq e$. Then we have

$$e \leq i \ \& \ e \leq n < m(s, e) \ \& \ p_i^m < y(s, n, e)$$
$$\& \ d(s, p_i^m) \neq d(s-1, p_i^m) .$$

It follows from the definition of $d(s, p_i^m)$ that $r(s, n, e) = 0$.

LEMMA 6. Let $y(s, n, e) > 0$, $m(s, e) > n > e$ and $r(s, n, e) = 1$. Then $m(s+1, e) > n$.

PROOF. Since $m(s, e) > n > e$, it follows from Case 1 of the definition of $m(s, e)$ and remark (R3) that

$$y(s, t, e) > 0$$

whenever $e \leq t \leq n$. But then by remark (R2) and Lemma 5,

$$y(s, t, e) = y(s+1, t, e)$$

whenever $e \leq t \leq n$. Since $y(s+1, e, e) = y(s, e, e) > 0$, Case 1 of the definition of $m(s+1, e)$ does not hold. If Case 2 of the definition of $m(s+1, e)$ holds, then

$$y\big(s+1, m(s+1, e), e\big) \neq y\big(s, m(s+1, e), e\big) ,$$

and consequently, $n < m(s+1, e)$. If Case 3 holds, then

$$m(s+1, e) \geq m(s, e) > n .$$

For each $e > 0$, we say e is stable if for all $n \geq e$, $\lim_s y(s, n, e)$ exists and is positive. If e is not the Gödel number of a system of equations, then $y(s, n, e) = 0$ for all s and n, and consequently, e is not stable. It follows there are infinitely many e which are not stable, since there are infinitely many e which are not Gödel numbers of sets of equations. We define:

$$e_0 = \mu e(e \text{ is not stable}) ;$$

$$e_{j+1} = \mu e(e > e_j \ \& \ e \text{ is not stable}) .$$

Thus $e_0 < e_1 < e_2 < \ldots$ is a listing of all the e which are not stable. For each j, let n_j be the least $n \geq e_j$ such that $\lim_s y(s, n, e_j)$ either does not exist or is equal to 0. Thus n_j is the least witness to the fact that e_j is not stable.

The combinatorial aspect of our reasoning is concentrated in the proof of Lemma 7.

LEMMA 7. For each k and v, there is an $s \geq v$ such that $(j)_{j<k}[m(s, e_j) \leq n_j \vee r(s, n_j, e_j) = 0 \vee y(s, n_j, e_j) = 0]$.

PROOF. Fix k and v. We suppose there is no s with the properties required by the conclusion of the lemma, and then show it is possible to define an infinite, descending sequence of natural numbers.

We propose the following system of equations as a means of defining two functions, $S(t)$ and $M(t)$, simultaneously by induction:

$$S(0) = v \ ;$$

$$M(t) = \mu j\Big[n_j < m\big(S(t), e_j\big) \ \& \ r\big(S(t), n_j, e_j\big) = 1 \ \& \ y\big(S(t), n_j, e_j\big) > 0\Big] \ ;$$

$$S(t+1) = \mu s(Em)\Big[s \geq S(t) \ \& \ m < y\big(S(t), n_{M(t)}, e_{M(t)}\big) \ \& \ d(s, m) \neq d\big(S(t) - 1, m\big)\Big] \ .$$

Clearly $S(0)$ is well-defined and $S(0) \geq v$. Suppose $t \geq 0$ and $S(t)$ is well-defined and $S(t) \geq v$. Then $M(t)$ is well-defined and $M(t) < k$, since we have supposed the lemma to be false. Thus

$$y\big(S(t), n_{M(t)}, e_{M(t)}\big) > 0 \ .$$

Since $e_{M(t)}$ is not stable, there must be an $s > S(t)$ such that

$$y\big(s, n_{M(t)}, e_{M(t)}\big) \neq y\big(S(t), n_{M(t)}, e_{M(t)}\big) \ .$$

It follows there is an $s > S(t)$ and an m such that

$$m < y\big(S(t), n_{M(t)}, e_{M(t)}\big) \ \& \ d(s-1, m) \neq d\big(S(t) - 1, m\big) \ .$$

But then $S(t+1)$ is well-defined and $S(t+1) \geq S(t) \geq v$.

For each $t \geq 0$, let

$$u(t) = \mu m\Big[d\big(S(t+1), m\big) \neq d\big(S(t) - 1, m\big)\Big] \ .$$

Note that

$$d\big(S(t+1), u(t)\big) \neq d\big(S(t+1)-1, u(t)\big)$$

for all t; this last is a consequence of the definitions of S and u. Now we show $u(t) < u(t-1)$ for all $t > 0$. Fix $t > 0$. Since we have

$$u(t) < y\big(S(t), n_{M(t)}, e_{M(t)}\big)$$

as a consequence of the definitions of $u(t)$ and $S(t+1)$, it will be sufficient to show $y\big(S(t), n_{M(t)}, e_{M(t)}\big) \leq u(t-1)$. Let

$$e = e_{M(t)}, \quad s = S(t) \quad \text{and} \quad n = n_{M(t)} \quad .$$

We know $d\big(s, u(t-1)\big) \neq d\big(s-1, u(t-1)\big)$. This last means there must be an i and an m such that either

$$u(t-1) = 5 \cdot 7^m \text{ and } m > 0 \quad ,$$

or

$$u(t-1) = p_i^m \quad .$$

We first suppose $u(t-1) = 5 \cdot 7^m$ and $m > 0$. Then $g(s) = m$; and $r(s, n, e) = 1$, since $M(t)$ is well-defined. But then it follows from the definition of $r(s, n, e)$ that $y(s, n, e) \leq 5 \cdot 7^{g(s)} = u(t-1)$.

Now we suppose $u(t-1) = p_i^m$. Again $r(s, n, e) = 1$. Suppose $i < e$. Then

$$i < e \leq n \,\&\, d(s, p_i^m) \neq d(s-1, p_i^m) \quad .$$

Since $r(s, n, e) = 1$, it follows $y(s, n, e) \leq p_i^m$. Suppose $i \geq e$. Then

$$e \leq i \,\&\, e \leq n < m(s, e) \,\&\, r(s, n, e) = 1 \,\&\, d(s, p_i^m) \neq d(s-1, p_i^m) \quad ,$$

since $M(t)$ is well-defined. But then it follows from the definition of $d(s, p_i^m)$ that $y(s, n, e) \leq p_i^m = u(t-1)$.

We introduce two predicates.

$A(e)$: if e is stable, then the set $\{m(s, e) | s \geq 0\}$ is finite.

$B(e)$: $\lim_n d(p_e^n)$ exists and is equal to $1 - c(e)$.

We will prove $(e)A(e)$ and $(e)B(e)$ by a simultaneous induction. It will follow from $(e)A(e)$ that $\underline{a} \nleq \underline{d}$. It will follow from $(e)B(e)$ that $\underline{c} \leq \underline{d}'$.

LEMMA 8. $(e)[(i)_{i<e}B(i) \rightarrow A(e)]$.

PROOF. Fix e. Suppose B(i) holds for all $i < e$. Part of the hypothesis of our theorem is: $\underline{a} \nleq \underline{g}$. We suppose A(e) is false and show $\underline{a} \leq \underline{g}$. Thus we have that $\lim_s y(s, n, e)$ exists and is positive for all $n \geq e$ and that the set $\{m(s, e) | s \geq 0\}$ is infinite. Let R(n, s) denote the predicate

$$(m)_{m>0}(t)[e \leq t \leq n \;\&\; 5 \cdot 7^m < y(s, t, e) \rightarrow d(s-1, 5 \cdot 7^m) = d(5 \cdot 7^m)]$$
$$\&\; (i)(m)(t)[i < e \leq t \leq n \;\&\; p_i^m < y(s, t, e) \rightarrow d(s-1, p_i^m) = d(p_i^m)]$$
$$\&\; m(s, e) > n.$$

We know B(i) holds for all $i < e$; this means $\lim_m d(p_i^m)$ exists for all $i < e$. For each $i < e$, let r(i) be a natural number with the property that

$$(m)[m \geq r(i) \rightarrow d(p_i^m) = d(p_i^{r(i)})] \quad .$$

We define a function z(n) recursive in the set G of degree $\underline{g}$ as follows: first we require that $z(n) = 1$ for all n not of the form p_i^m $(m \geq 0 \;\&\; i \geq 0)$ or $5 \cdot 7^m$ $(m > 0)$; then we specify:

$$z(p_i^m) = \begin{cases} d\left(p_i^{r(i)}\right) & \text{if } i < e \;\&\; m \geq r(i) \\ d(p_i^m) & \text{if } i < e \;\&\; m < r(i) \\ 1 & \text{otherwise ;} \end{cases}$$

$$z(5 \cdot 7^{m+1}) = \begin{cases} 0 & \text{if } m+1 \in G \\ 1 & \text{otherwise} \end{cases} \quad .$$

Since $d(5 \cdot 7^{m+1}) = 0$ if and only if $m + 1 \in G$, we can rewrite the predicate R(n, s) as follows:

$$(m)_{m>0}(t)[e \leq t \leq n \;\&\; 5 \cdot 7^m < y(s, t, e) \rightarrow d(s-1, 5 \cdot 7^m) = z(5 \cdot 7^m)]$$
$$\&\; (i)(m)(t)[i < e \leq t \leq n \;\&\; p_i^m < y(s, t, e) \rightarrow d(s-1, p_i^m) = z(p_i^m)]$$
$$\&\; m(s, e) > n.$$

It is now clear that the predicate R is recursive in the set G.

We claim $(n)(Es)R(n, s)$. Fix n. Since $\lim_s y(s, n, e)$ exists for all $n \geq e$, there is a y such that

$$(s)(t)[e \leq t \leq n \rightarrow y \geq y(s, t, e)]$$

Let s' be so large that

$$d(s - 1, w) = d(w)$$

for all $s \geq s'$ and $w < y$. Since the set $\{m(s, e) | s \geq 0\}$ is infinite, there is an $s \geq s'$ such that $m(s, e) > n$. But then $R(n, s)$ holds.

We define

$$w(n) = \mu s R(n, s) \quad ;$$

the function w is recursive in the set G. We now claim

$$y(w(n), n, e) = \lim_s y(s, n, e)$$

for all $n > e$. Fix $n > e$. Our claim is proved by an induction on $s \geq w(n)$. Fix $s \geq w(n)$ and suppose

$$y(w(n), n, e) = y(s, n, e) \ \& \ R(n, s) \quad .$$

Since $R(n, s)$ holds, we have $m(s, e) > n > e$; it follows from remark (R3) and Case 1 of the definition of $m(s, e)$ that

$$(t)[e \leq t \leq n \rightarrow y(s, t, e) > 0]$$

Since $R(n, s)$ holds, we have

$$\begin{aligned} &(m)_{m>0}(t)[e \leq t \leq n \ \& \ 5 \cdot 7^m < y(s, t, e) \\ &\qquad\qquad \rightarrow d(s - 1, 5 \cdot 7^m) = d(5 \cdot 7^m)] \\ &\& \ (i)(m)(t)[i < e \leq t \leq n \ \& \ p_i^m < y(s, t, e) \\ &\qquad\qquad \rightarrow d(s - 1, p_i^m) = d(p_i^m)] \quad . \end{aligned}$$

It immediately follows that

$$(t)[e \leq t \leq n \rightarrow r(s, t, e) = 1] \quad .$$

But then by Lemma 5, we have

$$(t)[e \leq t \leq n \rightarrow y(s, t, e) = y(s + 1, t, e)] \quad ,$$

and by Lemma 6, we have

$$m(s + 1, e) > n \quad .$$

Recall that if $d(s-1, w) = d(w)$, then $d(s, w) = d(w)$. Then

$$y(w(n), n, e) = y(s+1, n, e) \mathbin{\&} R(n, s+1) \quad .$$

Now we show that

$$a(n) = U(y(w(n), n, e))$$

for all $n > e$. It will then be clear $\underline{a} \leq \underline{g}$, since the function w is recursive in the set G. Fix n. We know:

$$\lim_s a(s, n) = a(n) \quad ;$$
$$\lim_s y(s, n, e) = y(w(n), n, e) \quad ;$$
the set $\{m(s, e) | s \geq 0\}$ is infinite.

It follows there must be an s with the property that $a(s, n) = a(n)$, $y(s, n, e) = y(w(n), n, e)$, $n < m(s, e)$ and $m(s-1, e) < m(s, e)$. Since $e < n < m(s, e)$ and $m(s-1, e) \leq m(s, e)$, Case 3 of the definition of $m(s, e)$ must hold. But then

$$a(n) = a(s, n) = U(y(s, n, e)) = U(y(w(n), n, e)) \quad ,$$

since $e < n < m(s, e)$ and $m(s-1, e) < m(s, e)$.

LEMMA 9. $(e)[(i)_{i \leq e} A(i) \rightarrow B(e)]$.

PROOF. Fix e. Suppose $A(i)$ holds for all $i \leq e$. We first suppose $c(e) = 0$ and show $\lim_n d(p_e^n)$ exists and is equal to 1. By Lemma 4, the set $\{h(s, e) | s \geq 0\}$ is finite; let h be its greatest member. Then by the definition of $d(s, p_e^n)$,

$$d(s, p_e^n) = d(s-1, p_e^n)$$

whenever $s > 0$ and $n \geq h$. But then

$$d(p_e^n) = \lim_s d(s, p_e^n) = d(0, p_e^n) = 1$$

for all $n \geq h$.

Now we suppose $c(e) = 1$ and show $\lim_n d(p_e^n)$ exists and is equal to 0. By Lemma 4, the set $\{h(s, e) | s \geq 0\}$ is infinite. If $i \leq e$ and i is stable, then the set $\{m(s, \overline{i}) | s \geq 0\}$ is finite, since $A(i)$ holds for all $i \leq e$. If $i \leq e$ and i is stable, let $m(i)$ be the greatest member of $\{m(s, \overline{i}) | s \geq 0\}$. If $i \leq e$ and i is not stable, let $m(i) = n_j$, where j is such that $i = e_j$. (Recall that e_j is the

j^{th} non-stable natural number.) If $i \leq e$ and $i \leq m < m(i)$, then $\lim_s y(s, m, i)$ exists. Let y be so large that

$$(s)(m)(i)[i \leq e \ \& \ i \leq m < m(i) \to y(s, m, i) \leq y] \quad .$$

We will show $d(p_e^n) = 0$ for all $n > y$. Fix $n > y$. Since the set $\{h(s, e) | s \geq 0\}$ is infinite, there is a v with the property that $h(v, e) > n$. Let k be such that if $i \leq e$ and i is not stable, then $i = e_j$ for some $j < k$. By Lemma 7, there is an $s \geq v$ such that

$$(j)_{j<k}[m(s, e_j) \leq n_j \vee r(s, n_j, e_j) = 0 \vee y(s, n_j, e_j) = 0] \quad .$$

If we can show

$$h(s, e) > n$$

and

$$(i)(m)[i \leq e \ \& \ i \leq m < m(s, i) \to r(s, m, i) = 0 \vee p_e^n \geq y(s, m, i)] \quad ,$$

then it will follow from the definition of $d(s, p_e^n)$ that $d(s, p_e^n) = 0$. We have $h(s, e) > n$, since $s \geq v$ and $h(v, e) > n$. Fix i and m and suppose $i \leq e$ and $i \leq m < m(s, i)$. Suppose i is stable. Then $m < m(i)$, since $m < m(s, i)$ and $m(i)$ is the greatest member of $\{m(s, i) | s \geq 0\}$. But then $y \geq y(s, m, i)$, and consequently, $p_e^n \geq y \geq y(s, m, i)$.

Now suppose i is not stable. Then $i = e_j$, where $j < k$, and $m(i) = n_j$. If $m < n_j$, then $m < m(i)$ and $p_e^n \geq y \geq y(s, m, i)$. Suppose $m \geq n_j$. Then $m(s, e_j) = m(s, i) > m \geq n_j$. This last means

$$r(s, n_j, e_j) = 0 \ \text{or} \ y(s, n_j, e_j) = 0 \quad .$$

If $r(s, n_j, i) = 0$, then by remark (R2), $r(s, m, i) = 0$, since $m \geq n_j$. Suppose $y(s, n_j, i) = 0$. Since $n_j < m(s, i)$, it follows from remark (R3) that $n_j = i$. But then $y(s, i, i) = 0$, and Case 1 of the definition of $m(s, i)$ holds. It follows $m(s, i) = i + 1$, $m = i$ and $y(s, m, i) = 0$.

LEMMA 10. $\underline{c} \leq \underline{d}'$.

PROOF. Observe that

$$(e)(Et)\Big[(m)_{m \geq t}\big(d(p_e^m) = 1\big) \vee (m)_{m \geq t}\big(d(p_e^m) = 0\big)\Big]$$

is an immediate consequence of $(e)B(e)$. We define

$$k(e) = \mu t\Big[(m)_{m \geq t}\big(d(p_e^m) = 1\big) \vee (m)_{m \geq t}\big(d(p_e^m) = 0\big)\Big]$$

The function k has degree less than or equal to $\underline{d}'$. It follows from $(e)B(e)$ that

$$c(e) = 1 - d\big(p_e^{k(e)}\big)$$

for all e.

LEMMA 11. $\underline{a} \nleq \underline{d}$.

PROOF. We suppose there is a Gödel number e such that

$$a(n) = \{e\}^d(n)$$

for all n, and show $A(e)$ is false. First we must show e is stable. Fix $n \geq e$; we wish to see that $\lim_s y(s, n, e)$ exists and is positive. Let

$$w = \mu y T_1^1\big(\tilde{d}(y), e, n, y\big)$$

Let s' be so large that $s' > w$ and $d(s, m) = d(m)$ whenever $s \geq s'$ and $m < w$. Then

$$y(s, n, e) = w \;\&\; U(w) = a(n)$$

for all $s > s'$; $w > 0$, since $U(0) = 0$ and $\dot{a}(n) > 0$.

Now we show the set $\{m(s, e) | s \geq 0\}$ is infinite. We fix $m > e$ and look for an s' with the property that $m(s', e) > m$. Let s be so large that $s > m$ and

$$a(s, t) = a(t) = \{e\}^d(t) = U\big(y(s, t, e)\big)$$

whenever $e \leq t \leq m$. If $m(s - 1, e) > m$, then $s - 1$ is the desired s'. Suppose $m(s - 1, e) \leq m$. Then

$$a(s, t) = U\big(y(s, t, e)\big)$$

whenever $e \leq t \leq m(s - 1, e)$; in addition, $y(s, e, e) > 0$, since $a(s, e) > 0$ and $U(0) = 0$. But then Case 3 of the definition of $m(s, e)$ must hold. Since $s > m$ and $a(s, t) = U\big(y(s, t, e)\big)$ whenever $e \leq t \leq m$, it follows that $m(s, e) > m$.

LEMMA 12. $\underline{d}' \leq \underline{c}$.

PROOF. We will define two functions, $E(e, n)$ and $L(e)$, simultaneously by induction on e so that each is recursive in the function $c(n)$

It will turn out that $E(e, e)$ is the representing function of the predicate

$$(Ey)T_1^1(\tilde{d}(y), e, e, y) \quad ,$$

and that, consequently, $\underline{d}' \leq \underline{c}$. We will combine the definitions of E and L with a proof by induction on e of

$$(e)(n)\Big\{E(e, n) = 0 \longleftrightarrow \Big[n \geq e \;\&\; (w)(Es)\big(s > w \;\&\; n < m(s, e)\big) \;\&\; (m)\big(n \geq m \geq e \rightarrow (Ey)T_1^1(\tilde{d}(y), e, n, y)\big)\Big]\Big\}$$

and

$$(e)(m)\Big[m \geq L(e) \rightarrow d(p_e^m) = d\big(p_e^{L(e)}\big)\Big] \quad .$$

Fix $e \geq 0$. Our induction hypothesis has two parts:

(1, e) for all $i < e$ and all n, $E(i, n)$ has been defined and

$$E(i, n) = 0 \longleftrightarrow \Big[n \geq i \;\&\; (w)(Es)\big(s > w \;\&\; n < m(s, i)\big) \;\&\; (m)\big(n \geq m \geq i \rightarrow (Ey)T_1^1(\tilde{d}(y), i, m, y)\big)\Big] \quad ;$$

(2, e) for all $i < e$, $L(i)$ has been defined and

$$(m)\Big[m \geq L(i) \rightarrow d(p_i^m) = d\big(p_i^{L(i)}\big)\Big] \quad .$$

We proceed to define $E(e, n)$ for all n, to verify $(1, e + 1)$, to define $L(e)$, and finally, to verify $(2, e + 1)$.

Let $(1, e + 1, n)$ denote the following predicate: for each $t < n$, $E(e, t)$ has been defined and

$$E(e, t) = 0 \longleftrightarrow \Big[t \geq e \;\&\; (w)(Es)\big(s > w \;\&\; t < m(s, e)\big) \;\&\; (m)\big(t \geq m \geq e \rightarrow (Ey)T_1^1(\tilde{d}(y), e, m, y)\big)\Big] \quad .$$

To verify $(1, e + 1)$, it will suffice to prove $(1, e + 1, n)$ for all n. Our plan is to define $E(e, n)$ and prove $(1, e + 1, n)$ for all n by induction on n. We begin by setting $E(e, t) = 1$ for all $t < e$. Then it is clear that $(1, e + 1, t)$ holds for all $t \leq e$. Fix $n \geq e$ and suppose $(1, e + 1, n)$ holds. We proceed to define $E(e, n)$ and prove $(1, e + 1, n + 1)$. The definition of $E(e, n)$ has two cases;

CASE 1. $(Et)\big(e \leq t < n \;\&\; E(e, t) \neq 0\big)$. We set $E(e, n) = 1$.

CASE 2. Otherwise. It follows from $(1, e + 1, n)$ that

$$(m)\Big(n > m \geq e \to (Ey)T_1^1\big(\tilde{d}(y),\ e,\ m,\ y\big)\Big) \quad .$$

For each m such that $n > m \geq e$, let

$$y(m) = \mu y T_1^1\big(\tilde{d}(y),\ e,\ m,\ y\big) \quad .$$

Let y^* be the least upper bound of the set $\{y(m) \mid n > m \geq e\}$. Let

$$s^* = \mu s\Big[s > y^* \ \&\ (i)\big(i < y^* \to d(s-1,\ i) = d(i)\big)\Big] \quad .$$

Recall that if $d(s-1,\ i) = d(i)$, then $d(s',\ i) = d(i)$ for all $s' \geq s$. It follows from the definition of $y(s,\ m,\ e)$ and from the fact that 0 is not the Gödel number of a deduction that

$$(s)(m)[s \geq s^* \ \&\ n > m \geq e \to y(s,\ m,\ e) = y(m) > 0] \quad .$$

We define

$$E(e,\ n) = \begin{cases} 0 & \text{if } (Es)\Big\{y(s,\ n,\ e) > 0 \ \&\ n < m(s,\ e) \ \&\ s > s^* \\ & \quad \&\ (t)_{t>0}\big(5 \cdot 7^t < y(s,\ n,\ e) \to d(s-1,\ 5 \cdot 7^t) = d(5 \cdot 7^t)\big) \\ & \quad \&\ (i)_{i<e}\Big[(m)\big(m < L(i) \to d(s-1,\ p_i^m) = d(p_i^m)\big) \\ & \quad \&\ (m)\Big(L(i) \leq m < y(s,\ n,\ e) \to d(s-1,\ p_i^m) = d\big(p_i^{L(i)}\big)\Big)\Big]\Big\} \\ 1 & \text{otherwise.} \end{cases}$$

The verification of $(1,\ e+1,\ n+1)$ has two parts. First we suppose $E(e,\ n) = 0$ and show

$$n \geq e \ \&\ (w)(Es)\big(s > w \ \&\ n < m(s,\ e)\big)$$
$$\&\ (m)\Big(n \geq m \geq e \to (Ey)T_1^1\big(\tilde{d}(y),\ e,\ m,\ y\big)\Big) \quad .$$

We know $n \geq e$. Since $E(e,\ n) = 0$, Case 2 of the definition of $E(e,\ n)$ must hold; let s be the natural number whose existence is required by the fact $E(e,\ n) = 0$. Thus

$$y(s,\ n,\ e) > 0 \ \&\ n < m(s,\ e) \ \&\ s > s^* \quad .$$

We prove $y(s',\ n,\ e) = y(s,\ n,\ e) \ \&\ n < m(s',\ e)$ for all $s' \geq s$ by an induction on s'. Fix $s' \geq s$ and suppose

$$y(s',\ n,\ e) = y(s,\ n,\ e) \ \&\ n < m(s',\ e) \quad .$$

We claim $d(s',\ j) = d(s'-1,\ j)$ for all $j < y(s',\ n,\ e)$. Suppose (for the sake of a reductio ad absurdum) that our claim is false. Then j must be of the form $5 \cdot 7^t$ $(t > 0)$ or of the form p_i^m $(m \geq 0)$. We know s

has the property that

$$(t)_{t>0}\left(5 \cdot 7^t < y(s, n, e) \rightarrow d(s-1, 5 \cdot 7^t) = d(5 \cdot 7^t)\right) .$$

Since $s' \geq s$ and $y(s', n, e) = y(s, n, e)$, it follows j cannot have the form $5 \cdot 7^t$ $(t > 0)$. This means there is an i and an m such that (0) and (1) hold:

$$(0) \quad d(s', p_i^m) \neq d(s'-1, p_i^m) \mathbin{\&} p_i^m < y(s', n, e) \quad ;$$

$$(1) \quad (i')_{i'<i}(m')[d(s', p_{i'}^{m'}) = d(s'-1, p_{i'}^{m'}) \vee p_{i'}^{m'} \geq y(s', n, e)] .$$

Note that $m < y(s, n, e)$, since $p_i^m < y(s', n, e) = y(s, n, e)$. Suppose $i < e$. Then s has the property that

$$d(s-1, p_i^m) = \begin{cases} d(p_i^m) & \text{if } m < L(i) \\ d(p_i^{L(i)}) & \text{if } m \geq L(i) \end{cases} .$$

It follows from (2, e) that $d(s-1, p_i^m) = d(p_i^m)$. But this last contradicts the first clause of (0), since $s' \geq s$. Thus we have shown $i \geq e$. We now have

$$e \leq i \mathbin{\&} e \leq n < m(s', e) \mathbin{\&} p_i^m < y(s', n, e)$$
$$\mathbin{\&} d(s', p_i^m) \neq d(s'-1, p_i^m) .$$

It follows from the definition of $d(s', p_i^m)$ that $r(s', n, e) = 0$. This last means there is an i', an m' and a t such that

$$i' < e \leq t \leq n \mathbin{\&} p_{i'}^{m'} < y(s', t, e) \mathbin{\&} d(s', p_{i'}^{m'}) \neq d(s'-1, p_{i'}^{m'}) .$$

If $t = n$, then (1) is false, since $i' < e \leq i$. So $n > t \geq e$. But then

$$p_{i'}^{m'} < y(s', t, e) = y(t) \leq y^* ,$$

since $s' \geq s > s^*$. But this is absurd, since $d(s'-1, p_{i'}^{m'}) = d(s', p_{i'}^{m'})$ is a consequence of $p_{i'}^{m'} < y^*$ and $s > s^*$.

That completes the proof of our claim that

$$(j)[j < y(s', n, e) \rightarrow d(s', j) = d(s'-1, j)] .$$

It follows immediately that

$$y(s'+1, n, e) = y(s', n, e) = y(s, n, e) > 0 .$$

Since $s' \geq s^*$, we now have

$$(m)[n \geq m \geq e \rightarrow y(s'+1, m, e) = y(s', m, e) > 0] .$$

This last will be enough to show $n < m(s' + 1, e)$. Case 1 of the definition of $m(s' + 1, e)$ cannot hold because $y(s' + 1, e, e) > 0$. If Case 2 of the definition of $m(s' + 1, e)$ holds, then

$$y\big(s' + 1, m(s' + 1, e), e\big) \neq y\big(s', m(s' + 1, e), e\big) \quad ,$$

and consequently, $n < m(s' + 1, e)$. If Case 3 holds, then $n < m(s', e) \leq m(s' + 1, e)$.

Thus we have shown $y(s', n, e) = y(s, n, e) > 0$ and $n < m(s', e)$ for all $s' \geq s$. It follows that

$$(w)(Es)\big(s > w \;\&\; n < m(s, e)\big) \;\&\; (Ey)T_1^1\big(\tilde{d}(y), e, n, y\big) \quad .$$

Since $E(e, n) = 0$, it follows from $(1, e + 1, n)$ that

$$(m)\Big[n > m \geq e \rightarrow (Ey)T_1^1\big(\tilde{d}(y), e, m, y\big)\Big] \quad .$$

We now perform the second half of the verification of $(1, e + 1, n + 1)$. We suppose

$$n \geq e \;\&\; (w)(Es)\big(s > w \;\&\; n < m(s, e)\big)$$
$$\&\; (m)\big(n \geq m > e \rightarrow (Ey)T_1^1\big(\tilde{d}(y), e, m, y\big)\big) \quad ,$$

and show $E(e, n) = 0$. It follows from $(1, e + 1, n)$ that

$$(t)[e \leq t < n \rightarrow E(e, t) = 0] \quad .$$

This last means Case 2 of the definition of $E(e, n)$ holds. Let

$$z = \mu y T_1^1\big(\tilde{d}(y), e, n, y\big) > 0 \quad .$$

Let w be so large that $w > z + s^*$ and

$$(j)(i)_{i<e}\Big[j < p_e^z \vee j < p_i^{L(i)} \rightarrow d(w - 1, j) = d(j)\Big] \quad .$$

Note that $y(s, n, e) = z > 0$ for all $s \geq w$. Let s be so large that $s > w$ and $n < m(s, e)$. With the help of $(2, e)$, it is readily seen that s has the properties needed to conclude $E(e, n) = 0$.

The definition of $L(e)$ has two cases.

CASE 1. $c(e) = 0$. By $B(e)$, $\lim_m d(p_e^m) = 1$. We set

$$L(e) = \mu t(s)(m)[m \geq t \rightarrow d(s, p_e^m) = 1] \quad .$$

CASE 2. $c(e) = 1$. It is a consequence of the definition of $y(s, n, i)$ that $(Ey)T_1^1\big(\tilde{d}(y), i, n, y\big) \;\&\; n \geq i \longleftrightarrow \lim_s y(s, n, i)$ exists and is positive. It follows from $(1, e+1)$ that

$$(i)_{i \le e}(n)\Big[E(i, n) = 0 \rightarrow (Es)\big(n < m(s, i)\big) \\ \& \lim_s y(s, n, i) \text{ exists and is positive}\Big] .$$

But then it follows from $(i)A(i)$ that

$$(i)_{i \le e}(Et)[t \ge i \ \& \ E(i, t) = 1] .$$

Let

$$t(i) = \mu t(t \ge i \ \& \ E(i, t) = 1)$$

for each $i \le e$. Then we have

$$(i)_{i \le e}(t)[i \le t < t(i) \rightarrow \lim_s y(s, t, i) \\ \text{exists and is positive}] .$$

We define $L(e)$ to be the least upper bound of the set

$$\{y(s, t, i) \mid i \le e \ \& \ i \le t < t(i) \ \& \ s \ge 0\} .$$

For each $i \le e$, it follows from $(1, e+1)$ and the definition of $t(i)$ that $t(i)$ satisfies either (2) or (3):

(2) $(Ew)(s)[s > w \rightarrow t(i) \ge m(s, i)]$;

(3) either $\lim_s y(s, t(i), i)$ does not exist or it equals zero.

We now verify $(2, e+1)$. We must show

$$(n)\Big[n \ge L(e) \rightarrow d(p_e^n) = d\big(p_e^{L(e)}\big)\Big] .$$

If $c(e) = 0$, then Case 1 of the definition of $L(e)$ holds, and there is nothing to show. Suppose $c(e) = 1$. Then by $B(e)$, $\lim_n d(p_e^n) = 0$. Fix $n \ge L(e)$. We must show $d(p_e^n) = 0$. The remainder of our argument resembles the second half of the argument of Lemma 9. Define v and k as in Lemma 9. Let w be so large that if $i \le e$ and $t(i)$ satisfies (2), then

$$(s)[s \ge w \rightarrow t(i) \ge m(s, i)] .$$

By Lemma 7, there is an $s \ge v + w$ such that

$$(j)_{j<k}[m(s, e_j) \le n_j \vee r(s, n_j, e_j) = 0 \vee y(s, n_j, e_j) = 0] .$$

If we can show

$$h(s, e) > n$$

and

$$(i)(m)[i \leq e \ \& \ i \leq m < m(s, i) \rightarrow r(s, m, i) = 0 \vee p_e^n \geq y(s, m, i)] \quad ,$$

then it will be clear $d(s, p_e^n) = 0$. We have $h(s, e) > n$, since $s \geq v$ and $h(v, e) > n$. Fix i and m and suppose $i \leq e$ and $i \leq m < m(s, i)$. Suppose i is such that (2) holds. Then $m < t(i)$, since $m < m(s, i)$ and $s \geq w$. But then $L(e) \geq y(s, m, i)$, and consequently, $p_e^n \geq y(s, m, i)$.

Suppose i is such that (3) holds. Then i is not stable, and there is a $j < k$ such that $e_j = i$; in addition, $t(i) = n_j$. If $m < n_j$, then $m < t(i)$ and $p_e^n \geq L(e) \geq y(s, m, i)$. Suppose $m \geq n_j$. Then $m(s, e_j) = m(s, i) > m \geq n_j$. This last means

$$r(s, n_j, e_j) = 0 \quad \text{or} \quad y(s, n_j, e_j) = 0 \quad .$$

If $r(s, n_j, e_j) = 0$, then by remark (R2), $r(s, m, i) = 0$, since $m \geq n_j$. Suppose $y(s, n_j, e_j) = 0$. Since $n_j < m(s, i)$, it follows from remark (R3) that $n_j = i$. But then $y(s, i, i) = 0$, $m(s, i) = i + 1$, $m = i$ and $y(s, m, i) = 0$.

We still have to show E and L are recursive in c. It would be impractical to exhibit a system of equations which define E and L recursively in c; we content ourselves by indicating intuitively how such a system may be obtained. Fix $e \geq 0$. We have $E(e, t) = 1$ for all $t < e$. Fix $n \geq e$. The definition of $E(e, n)$ has two cases; the determination of which case applies is made in an effective manner from the values of $E(e, t)(t \leq n)$. If Case 1 applies, $E(e, n) = 1$. Suppose Case 2 applies. Then y^* is defined with the help of a predicate of degree $\underline{d}$; s^* is obtained from the values of $d(i)(i < y^*)$ with the aid of a recursive predicate. Then the value of $E(e, n)$ is determined from the values of s^* and $d(p_i^m)(i < e$ and $m \leq L(i))$ by means of a predicate of degree $\underline{g}'$ (g tells us the value of $d(5 \cdot 7^t)$). Thus the value of $E(e, n)$ can be expressed in terms of the values of $E(e, t)(t < n)$ and $L(i)(i < e)$ with the help of a predicate of degree $\underline{d} \cup \underline{g}'$. Similarly, the value of $L(e)$ can be expressed in terms of the values of $E(i, t)(i \leq e$ and $t \geq 0)$ with the help of a predicate of degree $\underline{c} \cup \underline{0}'$. But then the functions E and

L are recursive in c, since

$$\underline{d} \cup \underline{g}' \cup \underline{c} \cup \underline{0}' = \underline{c} .$$

Since we have shown

$$(e)(n)\Big\{E(e, n) = 0 \longleftrightarrow \Big[n \geq e \ \& \ (w)(Es)\big(s > w \ \& \ n < m(s, e)\big) \\ \& \ (m)\big(n \geq m \geq e \rightarrow (Ey)T_1^1(\tilde{d}(y), e, m, y)\big)\Big]\Big\} ,$$

it follows from remark (R1) that

$$(e)\Big[E(e, e) = 0 \longleftrightarrow (Ey)T_1^1\big(\tilde{d}(y), e, e, y\big)\Big] ,$$

and that consequently, $\underline{d}' \leq \underline{c}$.

When $\underline{b} > \underline{0}$, the changes needed in the above argument are largely notational. Let b be a function of degree $\underline{b}$. The notion of recursiveness is replaced throughout by the notion of recursiveness in b. The functions a(s, n) f(s, n) and d(s, n) are now recursive in b. The function $\lim_s d(s, n)$ is now the representing function of a set recursively enumerable in b. Lemmas 4-11 are unchanged.

COROLLARY 1. If $\underline{b}$ and $\underline{c}$ are degrees, the following conditions are equivalent:

(i) $\underline{b}' \leq \underline{c} \leq \underline{b}''$ and $\underline{c}$ is recursively enumerable in $\underline{b}'$;

(ii) there is a $\underline{d}$ such that $\underline{b} \leq \underline{d} \leq \underline{b}'$ and $\underline{d}' = \underline{c}$;

(iii) there is a $\underline{d}$ such that $\underline{b} \leq \underline{d} \leq \underline{b}'$, $\underline{d}$ is recursively enumerable in $\underline{b}$, and $\underline{d}' = \underline{c}$.

PROOF. It is clear that (iii) → (ii) and (ii) → (i). To prove (i) → (iii), apply Theorem 3 with $\underline{a} = \underline{b}'$ and $\underline{g} = \underline{b}$.

Shoenfield [23] proved the equivalence of (i) and (ii); his argument made use of the priority method in a form similar to that of Section 4. Each of his "requirements" was "injured" only finitely often. We obtained a stronger result by permitting each "requirement" to be "injured" infinitely often. Our extension of the priority method provides stronger results than the priority method of Friedberg and Muchnik for reasons similar to the reasons why the latter method provides stronger results than the diagonal method of Kleene and Post [9].

For each degree $\underline{b}$, let $R(\underline{b})$ denote the set of all degrees greater than or equal to $\underline{b}$, recursively enumerable in $\underline{b}$ and less than or equal to $\underline{b}'$. Shoenfield [23] has shown that there is a degree between $\underline{b}$ and $\underline{b}'$ which is not a member of $R(\underline{b})$; we prove his result by a different method in Section 9. Let j denote the jump operator: $j(\underline{b}) = \underline{b}'$. Then Corollary 1 tells us that j maps $R(\underline{b})$ onto $R(\underline{b}')$; this last fact may be expressed as follows:

$$jR = Rj$$

Thus we have obtained a commutativity law concerning quantification.

Let $\underline{c}$ be a member of $R(\underline{b}')$. It is immediate from Theorems 2 and 3 that $j^{-1}(\underline{c}) \cap R(\underline{b})$ has more than one member. It can easily be shown, using the methods of Theorem 2 and 3 that $\underline{c}$ has infinitely many pre-images under j in $R(\underline{b})$. Thus j is an order-preserving map of $R(\underline{b})$ onto $R(\underline{b}')$ which is not one-one.

COROLLARY 2. Let $\underline{b}$ and $\underline{c}$ be degrees and n be a natural number such that $\underline{b}^{(n)} \leq \underline{c}$ and $\underline{c}$ is recursively enumerable in $\underline{b}^{(n)}$. Then there is a degree $\underline{d}$ recursively enumerable in $\underline{b}$ and $\geq \underline{b}$ such that $\underline{d}^{(n)} = \underline{c}$.

PROOF. Let $\underline{h}_0 = \underline{c}$. By making n consecutive applications of Corollary 1, we obtain a finite sequence of degrees, $\underline{h}_1, \underline{h}_2, \ldots, \underline{h}_n$, such that when $1 \leq i \leq n$,

$$\underline{b}^{(n-i)} \leq \underline{h}_i \leq \underline{b}^{(n-i+1)}, \ \underline{h}_i' = \underline{h}_{i-1}$$

and $\underline{h}_i$ is recursively enumerable in $\underline{b}^{(n-i)}$. Let $\underline{d} = \underline{h}_n$. Then d is recursively enumerable in $\underline{b}$ and $\underline{d}^{(n)} = \underline{h}_n^{(n)} = \underline{h}_{n-1}^{(n-1)} = \underline{h}_0 = \underline{c}$.

We saw in Section 4 that for any degree $\underline{b}$ and natural number n there is a degree $\underline{c}$ such that $\underline{b}^{(n)} < \underline{c} < \underline{b}^{(n+1)}$ and $\underline{c}$ is recursively enumerable in $\underline{b}^{(n)}$. It follows from Corollary 2 that there is a degree $\underline{d}$ recursively enumerable in $\underline{b}$ such that

$$\underline{b} < \underline{d} < \underline{b}' < \underline{d}' < \underline{b}'' < \ldots < \underline{b}^{(n)} < \underline{d}^{(n)} < \underline{b}^{(n+1)}$$

In [23] Shoenfield proved there is a degree $\underline{d}$ (not necessarily recursively enumerable in $\underline{b}$) such that $\underline{b} < \underline{d} < \underline{b}' < \underline{d}' < \underline{b}''$; Shoenfield's result for the case $\underline{b} = \underline{0}$ was announced without proof by Friedberg in [4]. We do not know if there exists a degree $\underline{d}$ such that

$$\underline{b}^{(n)} < \underline{d}^{(n)} < \underline{h}^{(n+1)}$$

for all $n \geq 0$. If such a $\underline{d}$ exists, then by Corollary 1, $\underline{d}$ can be given the additional property of recursive enumerability in $\underline{b}$.

Corollary 3 follows from Corollary 2 and is a "constructivization" of Theorem 1.

COROLLARY 3. A degree $\underline{c}$ is the completion of a recursively enumerable degree if and only if $\underline{c} \geq \underline{0}'$ and $\underline{c}$ is recursively enumerable in $\underline{0}'$.

COROLLARY 4. There exists a recursively enumerable degree $\underline{d}$ such that $\underline{d} < \underline{0}' < \underline{0}'' = \underline{d}'$.

PROOF. Let $\underline{b} = \underline{g} = \underline{0}$, $\underline{c} = \underline{0}''$ and $\underline{a} = \underline{0}'$, and then apply Theorem 3. Then $\underline{d}$ is recursively enumerable, $\underline{0}' \nleq \underline{d}$ and $\underline{d}' = \underline{0}''$.

COROLLARY 5. If $\underline{g}$ is a recursively enumerable degree such that $\underline{g}' < \underline{0}''$, then there exists a recursively enumerable degree $\underline{d}$ such that $\underline{g} < \underline{d} < \underline{0}'$ and $\underline{d}' = \underline{0}''$.

PROOF. Let $\underline{b} = \underline{0}$, $\underline{c} = \underline{0}''$ and $\underline{a} = \underline{0}'$. We have $\underline{0}' \nleq \underline{g}$, since $\underline{g}' < \underline{0}''$. Apply Theorem 3. Then $\underline{d}$ is recursively enumerable, $\underline{g} \leq \underline{d}$, $\underline{0}' \nleq \underline{d}$ and $\underline{d}' = \underline{0}''$. We have $\underline{g} < \underline{d}$, since $\underline{g}' < \underline{d}' = \underline{0}''$.

We are now able to describe completely the possible effects of the jump operator on the ordering of the recursively enumerable degrees. For each of the following nine statements there exist recursively enumerable degrees with the properties indicated:

(J1) $\underline{a}' \cup \underline{b}' < (\underline{a} \cup \underline{b})'$;

(J2) $\underline{0} < \underline{a} < \underline{0}'$ & $\underline{0}' = \underline{a}'$;

(J3) $\underline{0} < \underline{a} < \underline{0}'$ & $\underline{0}' < \underline{a}' < \underline{0}''$;

(J4) $\underline{0} < \underline{a} < \underline{0}'$ & $\underline{a}' = \underline{0}''$;

(J5) $\underline{a} < \underline{b}$ & $\underline{a}' = \underline{b}'$;

(J6) $\underline{a} < \underline{b}$ & $\underline{a}' < \underline{b}'$;

(J7) $\underline{a} \mid \underline{b}$ & $\underline{a}' = \underline{b}'$;

(J8) $\underline{a} \mid \underline{b}$ & $\underline{a}' < \underline{b}'$;

(J9) $\underline{a} \mid \underline{b}$ & $\underline{a}' \mid \underline{b}'$.

Proof of (J1) and (J7): by Corollary 1 of Theorem 2, there exist recursively enumerable degrees $\underline{a}$ and $\underline{b}$ such that $\underline{a} \cup \underline{b} = \underline{0}' = \underline{a}' \cup \underline{b}'$; then $\underline{a}' \cup \underline{b}' < (\underline{a} \cup \underline{b})'$, $\underline{a}$ is incomparable with $\underline{b}$ and $\underline{a}'$ is equal to $\underline{b}'$. (J2) follows from Corollary 2 of Theorem 2. (J3) was proved in the remarks following Corollary 2 of Theorem 3. (J4) is Corollary 4 of Theorem 3. (J5) follows from (J2), and (J6) follows from (J3). Proof of (J8): by (J2) we have a recursively enumerable degree $\underline{a}$ such that $\underline{0} < \underline{a}$ and $\underline{a}' = \underline{0}'$; apply Theorem 3 to obtain a recursively enumerable degree $\underline{b}$ such that $\underline{b}' = \underline{0}''$ and $\underline{a} \nleq \underline{b}$; since $\underline{b}' \nleq \underline{a}'$, we have $\underline{b} \nleq \underline{a}$. Proof of (J9): by Section 4, there exist degrees $\underline{c}$ and $\underline{d}$ such that $\underline{0}' \leq \underline{c}$, $\underline{0}' \leq \underline{d}$, $\underline{c}$ and $\underline{d}$ are recursively enumerable in $\underline{0}'$, and $\underline{c}$ is incomparable with $\underline{d}$; apply Theorem 3 to obtain recursively enumerable degrees $\underline{a}$ and $\underline{b}$ such that $\underline{a}' = \underline{c}$ and $\underline{b}' = \underline{d}$; then $\underline{a}$ is incomparable with $\underline{b}$, since $\underline{a}'$ is incomparable with $\underline{b}'$.

In Section 2 we obtained some results concerning the non-existence of least upper bounds for infinite, ascending sequences of degrees. We continue this investigation with the help of the methods of Theorem 3.

THEOREM 4. Let $\underline{a}_0 < \underline{a}_1 < \underline{a}_2 < \ldots$ be an infinite, ascending sequence of simultaneously recursively enumerable degrees. Then there exists a recursively enumerable degree $\underline{d}$ such that $\underline{a}_0 < \underline{a}_1 < \underline{a}_2 < \ldots < \underline{d} < \underline{0}'$.

PROOF. Let $g(i, s)$ be a recursive function such that for each i, the set

$$A_i = \{g(i, s) \mid s \geq 0\}$$

has degree $\underline{a}_i$; we assume g is everywhere positive. Let f be a recursive function whose range is a set A of degree $\underline{0}'$. For each s and n, let

$$a(s, n) = \begin{cases} 2 & \text{if } (Ek)(k \leq s \;\&\; f(k) = n) \\ 1 & \text{otherwise} \end{cases} .$$

The function $a(s, n)$ is recursive. For each n, $\lim_s a(s, n)$ exists. Let $a(n)$ be the representing function of A. Then $\lim_s a(s, n) = 2 - a(n)$.

We will define four recursive functions, $y(s, n, e)$, $m(s, e)$, $r(s, n, e)$ and $d(s, n)$, simultaneously by induction on s.

Stage $s = 0$. We set $y(0, n, e) = m(0, e) = 0$ and $r(0, n, e) = d(0, n) = 1$ for all n and e.

Stage $s > 0$. We define $y(s, n, e)$ for all n and e:

$$y(s, n, e) = \begin{cases} \mu y T_1^1\Big(\prod_{i<y} p_i^{d(s-1,i)}, e, n, y\Big) \\ \qquad \text{if } (Ey)_{y \le s}\, T_1^1\Big(\prod_{i<y} p_i^{d(s-1,i)}, e, n, y\Big) \\ 0 \quad \text{otherwise} \end{cases}.$$

The definition of $m(s, e)$ has two cases:

CASE 1. There is an $n < m(s - 1, e)$ such that

$$a(s, n) \neq U\big(y(s, n, e)\big) \;\&\; y(s, n, e) \neq y(s - 1, n, e).$$

We set $m(s, e)$ equal to the least such n.

CASE 2. Otherwise. We set $m(s, e)$ equal to

$$\mu n\Big[m(s - 1, e) \le n < 2m(s - 1, e) + s \;\&\; (Et)_{t \le n}\big(a(s, t) \neq U\big(y(s, t, e)\big)\big)\Big].$$

Note that the least number operator in Case 2 is bounded.

We define $r(s, n, e)$ and $d(s, p_e^n)$ for all e and n by means of a simultaneous induction on e. Suppose $e \ge 0$ and $r(s, n, i)$ and $d(s, p_i^n)$ have been defined for all $i < e$ and all n; we define $r(s, n, e)$ and $d(s, p_e^n)$ for all n:

$$r(s, n, e) = \begin{cases} 0 \quad \text{if } (Ei)(Em)(Et)[i < e \;\&\; t \le n \\ \qquad \;\&\; p_i^m < y(s, t, e) \\ \qquad \;\&\; d(s, p_i^m) \neq d(s - 1, p_i^m)] \\ 1 \quad \text{otherwise} \end{cases};$$

$$d(s, p_e^n) = \begin{cases} d(s - 1, p_e^n) \quad \text{if } (v)_{v \le s}(g(e, v) \neq n) \\ d(s - 1, p_e^n) \quad \text{if } (Ei)(Et)[i \le e \\ \qquad \;\&\; t < m(s, i) \;\&\; p_e^n < y(s, t, i) \\ \qquad \;\&\; r(s, t, i) = 1] \\ 0 \quad \text{otherwise} \end{cases}.$$

We complete the construction by setting $d(s, m) = d(s - 1, m)$ for all m not a power of a prime.

Each of the four functions defined is recursive, because each of the functions $g(i, s)$ and $a(s, n)$ is recursive, and because at stage $s > 0$, all quantifiers and applications of the least number operator are bounded. For each n, let

$$d(n) = \lim_s d(s, n) \quad ;$$

thus $d(n) = 0$ if and only if there is an s such that $d(s, n) = 0$. Let $\underline{d}$ be the degree of the recursively enumerable set whose representing function is d.

LEMMA 13. If $n < m(s, e)$, then $y(s, n, e) > 0$.

PROOF. By induction on s. If $s = 0$, there is nothing to prove, since $m(0, e) = 0$ for all e. Fix $s > 0$ and suppose

$$(n)(e)[n < m(s - 1, e) \rightarrow y(s - 1, n, e) > 0] \quad .$$

Fix n and e and suppose $n < m(s, e)$. It follows from the definition of $m(s, e)$ that

$$a(s, n) = U\big(y(s, n, e)\big)$$

or

$$y(s, n, e) = y(s - 1, n, e) \quad .$$

Suppose $a(s, n) = U\big(y(s, n, e)\big)$. Then $y(s, n, e) > 0$, since $a(s, n) \geq 1$ and $U(0) = 0$. Suppose $y(s, n, e) = y(s - 1, n, e)$. If $n < m(s - 1, e)$, then $y(s - 1, n, e) > 0$. Suppose $n \geq m(s - 1, e)$. Then Case 2 of the definition of $m(s, e)$ applies, and consequently, $a(s, n) = U\big(y(s, n, e)\big)$, since $n < m(s, e)$.

LEMMA 14. If $n < m(s, e)$ and $r(s, n, e) = 1$, then $y(s + 1, n, e) = y(s, n, e)$.

PROOF. Since $n < m(s, e)$, it follows from Lemma 13 that $y(s, n, e) > 0$. But then

$$y(s, n, e) = \mu y T_1^1\Big(\prod_{i < y} p_i^{d(s-1,i)}, e, n, y\Big) \quad .$$

To show $y(s + 1, n, e) = y(s, n, e)$, it is sufficient to show

$$d(s - 1, i) = d(s, i)$$

for all $i < y(s, n, e)$. Suppose otherwise; then there must be an i and an m such that

$$d(s, p_i^m) \neq d(s - 1, p_i^m) \ \& \ p_i^m < y(s, n, e) \quad .$$

Since $r(s, n, e) = 1$, it follows $i \geq e$. But then we have

$$e \leq i \ \& \ n < m(s, e) \ \& \ p_i^m < y(s, n, e) \ \& \ r(s, n, e) = 1 \quad .$$

This last means $d(s, p_i^m) = d(s - 1, p_i^m)$.

LEMMA 15. If $n < m(s, e)$ and $r(s, n, e) = 1$, then $n < m(s + 1, e)$.

PROOF. Since $r(s, n, e) = 1$, it must be that

$$r(s, t, e) = 1$$

for all $t \leq n$. But then by Lemma 14,

$$y(s, t, e) = y(s + 1, t, e)$$

for all $t \leq n$. If Case 1 of the definition of $m(s + 1, e)$ holds, then $n < m(s + 1, e)$. If Case 2 holds, then $n < m(s, e) \leq m(s + 1, e)$.

For each $e > 0$, we say e is stable if for all n, $\lim_s y(s, n, e)$ exists and is positive. If e is not the Gödel number of a set of equations, then $y(s, n, e) = 0$ for all n and s. It follows there are infinitely many e which are not stable. We define:

$$e_0 = \mu e(e \text{ is not stable}) \quad ;$$

$$e_{j+1} = \mu e(e > e_j \text{ and } e \text{ is not stable}) \quad .$$

For each $j < k$, let n_j be the least n such that $\lim_s y(s, n, e)$ either does not exist or is equal to 0.

LEMMA 16. For each k and v, there is an $s \geq v$ such that $(j)_{j<k}[m(s, e_j) \leq n_j \vee r(s, n_j, e_j) = 0 \vee y(s, n_j, e_j) = 0]$.

PROOF. Our argument is almost identical with that of Lemma 7. We suppose there is no s with the required properties, and then show it is possible to define an infinite, descending sequence of natural numbers.

We define the functions $S(t)$, $M(t)$ and $u(t)$ in precisely the same fashion we did in Lemma 7. Fix $t > 0$. Let $e = e_{M(t)}$, $s = S(t)$ and $n = n_{M(t)}$. It follows by exactly the same arguments as in Lemma 7 that

$$u(t) < y(s, n, e), \; n < m(s, e), \; r(s, n, e) = 1$$

and

$$d(s, u(t - 1)) \neq d(s - 1, u(t - 1)) \quad .$$

We show $u(t) < u(t - 1)$ by showing $u(t - 1) \geq y(s, n, e)$. Since $d(s, w) = d(s - 1, w)$ for all w not a power of a prime, there must be an i and an m such that $u(t - 1) = p_i^m$.

First suppose $i < e$. Then we have

$$i < e \;\&\; d(s, p_i^m) \neq d(s - 1, p_i^m) \;\&\; r(s, n, e) = 1 \quad .$$

It follows from the definition of $r(s, n, e)$ that $p_i^m \geq y(s, n, e)$.

Now suppose $i \geq e$. Then we have

$$e \leq i \;\&\; n < m(s, e) \;\&\; r(s, n, e) = 1 \;\&\; d(s, p_i^m) \neq d(s - 1, p_i^m) \quad .$$

It follows from the definition of $d(s, p_i^m)$ that $p_i^m \geq y(s, n, e)$.

We introduce two predicates.

$A(e)$: if e is stable, then the set $\{m(s, e) \mid s \geq 0\}$ is finite.

$B(e)$: $(Et)(n)_{n \geq t}[d(p_e^n) = 0 \leftrightarrow n \in A_e]$.

We will prove $(e)A(e)$ and $(e)B(e)$ by a simultaneous induction. It will follow from $(e)A(e)$ that $\underline{d} < \underline{0}'$. It will follow from $(e)B(e)$ that $\underline{a}_e \leq \underline{d}$ for all e.

LEMMA 17. $(e)[(i)_{i < e}B(i) \rightarrow A(e)]$.

PROOF. Fix e. We know $\underline{a}_e < \underline{0}'$. We suppose $A(e)$ is false and show $\underline{0}' \leq \underline{a}_e$. Thus we have that $\lim_s y(s, n, e)$ exists and is positive for all n, and that the set $\{m(s, e) \mid s \geq 0\}$ is infinite. For each $i < e$ there is a t such that

$$(m)_{m \geq t}[d(p_i^m) = 0 \leftrightarrow n \in A_i] \quad ,$$

since $B(i)$ is true; for each $i < e$, let $t(i)$ be the least such t. We define a function $z(n)$ as follows: first we require that $z(n) = 1$

for all n not a power of a prime; then we specify

$$z(p_i^m) = \begin{cases} d(p_i^m) & \text{if } i < e \text{ \& } m < t(i) \\ 0 & \text{if } i < e \text{ \& } m \geq t(i) \text{ \& } m \in A_i \\ 1 & \text{otherwise.} \end{cases}$$

Since $\underline{a}_i \leq \underline{a}_e$ for every $i < e$, it follows $z(n)$ is recursive in A_e.

Let $R(n, s)$ denote the predicate

$$n < m(s, e) \text{ \& } (i)(m)(t)[i < e \text{ \& } t \leq n \text{ \& } p_i^m < y(s, t, e) \rightarrow d(s - 1, p_i^m) = d(p_i^m)] \quad .$$

The predicate $R(n, s)$ is recursive in the set A_e, because

$$d(p_i^m) = z(p_i^m)$$

whenever $i < e$.

We claim that $(n)(Es)R(n, s)$. Fix n. Since $\lim_s y(s, t, e)$ exists for all t, there is a y with the property that

$$y > y(s, t, e) \quad ,$$

for all s and all $t \leq n$. Let s' be so large that $d(s' - 1, w) = d(w)$ for all $w < y$. Since $\{m(s, e) | s \geq 0\}$ is infinite there is an $s \geq s'$ such that $n < m(s, e)$. But then we have $R(n, s)$.

Let $w(n)$ denote the function $\mu sR(n, s)$; clearly w is recursive in A_e.

Now we show $y(w(n), n, e) = \lim_s y(s, n, e)$ for all n. Again fix n. We show by induction on s that $y(w(n), n, e) = y(s, n, e)$ for all $s \geq w(n)$. Fix s and suppose $s \geq w(n)$ and

$$y(w(n), n, e) = y(s, n, e) \text{ \& } R(n, s) \quad .$$

Since $R(n, s)$ holds, we have $r(s, t, e) = 1$ for all $t \leq n$, and $n < m(s, e)$. It follows from Lemmas 14 and 15 that

$$y(s + 1, t, e) = y(s, t, e) \text{ \& } n < m(s + 1, e) \quad ,$$

for all $t \leq n$. But then $y(w(n), n, e) = y(s + 1, n, e)$ and $R(n, s + 1)$ holds.

Finally we show

$$a(n) = U\big(y(w(n), n, e)\big)$$

for all n; it will then be clear $\underline{0}' \leq \underline{a}_e$, since the function $a(n)$ is of degree $\underline{0}'$. Fix n, We know:

$$\lim_s a(s, n) = a(n) \quad ;$$

$$\lim_s y(s, n, e) = y(w(n), n, e) \quad ;$$

the set $\{m(s, e) | s \geq 0\}$ is infinite .

It follows there is an s such that $a(s, n) = a(n)$, $y(w(n), n, e) = y(s, n, e)$, $n < m(s, e)$ and $m(s - 1, e) < m(s, e)$. Since $m(s - 1, e) < m(s, e)$, Case 2 of the definition of $m(s, e)$ must hold. Since $n < m(s, e)$ and $m(s - 1, e) < m(s, e)$, we have

$$a(n) = a(s, n) = U\big(y(s, n, e)\big) = U\big(y(w(n), n, e)\big) \quad .$$

LEMMA 18. $(e)[(i)_{i \leq e} A(i) \rightarrow B(e)]$.

PROOF. Fix e. If $n \notin A_e$, then $d(s, p_e^n) = 1$ for all s, and consequently, $d(p_e^n) = 1$. Thus to prove $B(e)$, we must show

$$(Et)(n)_{n \geq t}[n \in A_e \rightarrow d(p_e^n) = 0] \quad .$$

If $i \leq e$ and i is stable, then it follows from $A(i)$ that the set $\{m(s, i) | s \geq 0\}$ is finite. For each stable $i \leq e$, let $m(i)$ be the greatest member of $\{m(s, i) | s \geq 0\}$. For each $i \leq e$ which is not stable, let $m(i) = n_j$, where j is such that $i = e_j$. Let t be so large that

$$(s)(m)(i)[i \leq e \ \& \ m < m(i) \rightarrow y(s, m, i) \leq t] \quad .$$

Fix $n \geq t$ and suppose $n \in A_e$. We show $d(p_e^n) = 0$. It will suffice to find an s such that $d(s, p_e^n) = 0$. Since $n \in A_e$, there is a v with the property that $g(e, v) = n$. By Lemma 16, there is an $s \geq v$ such that

$$(j)_{j \leq e}[m(s, e_j) \leq n_j \vee r(s, n_j, e_j) = 0 \vee y(s, n_j, e_j) = 0] \quad .$$

We will show

$$(i)(m)[i \leq e \ \& \ m < m(s, i) \rightarrow r(s, m, i) = 0 \vee p_e^n \geq y(s, m, i)] \quad .$$

It will then be clear that $d(s, p_e^n) = 0$.

Fix i and m so that $i \leq e$ and $m < m(s, i)$. Suppose i is stable. Then $m < m(i)$, since $m(s, i) \leq m(i)$. But then $y(s, m, i) \leq t$, and consequently, $y(s, m, i) \leq p_e^n$.

Now suppose i is not stable. Then $i = e_j$, where $j \leq i \leq e$, and $m(i) = n_j$. If $m < n_j$, then $m < m(i)$ and $p_e^n \geq y(s, m, i)$. Suppose $m \geq n_j$. Then $m(s, e_j) = m(s, i) > m \geq n_j$. This last means

$$r(s, n_j, e_j) = 0 \quad \text{or} \quad y(s, n_j, e_j) = 0 \quad .$$

If $r(s, n_j, i) = 0$, then $r(s, m, i) = 0$, since $m \geq n_j$. Suppose $y(s, n_j, i) = 0$. Then by Lemma 13, $n_j \geq m(s, i)$. This last is absurd, since $m \geq n_j$ and $m(s, i) > m$.

That concludes the proof of $(e)A(e)$ and $(e)B(e)$. $B(e)$ tells us that $\underline{a}_e \leq \underline{d}$. We suppose $\underline{0}' \leq \underline{d}$ and show $A(e)$ to be false for some e. Since the function $a(n)$ is of degree $\underline{0}'$, there must be a Gödel number e with the property that

$$a(n) = \{e\}^d(n)$$

for all n. First we show the set $\{m(s, e) | s \geq 0\}$ is infinite. We fix m and find an s such that $m(s, e) \geq m$. Let s' be so large that $s' \geq m$ and

$$a(s, t) = a(t) = \{e\}^d(t) = U\big(y(s, t, e)\big) \quad ,$$

for all $t \leq m$. If $m(s'-1, e) \geq m$, then $s'-1$ is the desired s. Suppose $m(s'-1, e) < m$. Then

$$a(s', t) = U\big(y(s', t, e)\big)$$

for all $t < m(s'-1, e)$, and consequently, Case 2 of the definition of $m(s', e)$ holds. But then $m(s', e) \geq m$, since $s' > m$; and s' is the desired s.

It remains only to see e is stable. Fix n and let $y = \{e\}^d(n)$. Let s be so large that $s \geq y$ and

$$d(s-1, i) = d(i)$$

for all $i < y$. Then $y(s', n, e) = y(s, n, e) = y > 0$ for all $s' \geq s$.

The proof of Theorem 4 requires only trivial modifications in order to become a proof of the following result:

Let A and B be sequences of simultaneously recursively enumerable degrees such that no member of B is less than or equal to any finite union of members of A; then there is a recursively enumerable degree $\underline{d}$ which is greater than or equal to every member of A but which is not greater than or equal to any member of B.

This last result may be viewed as a constructive version of the principal lemma of Section 2.

§7. AN INTERPOLATION THEOREM FOR RECURSIVELY ENUMERABLE DEGREES

Corollary 5 of Theorem 3 of Section 6 states that if $\underline{g}$ is a recursively enumerable degree such that $\underline{g}' < 0''$, then there exists a recursively enumerable degree $\underline{d}$ such that $\underline{g} < \underline{d} < \underline{0}'$ and $\underline{d}' = \underline{0}''$. The purpose of the present section is to extend this result. We assume complete familiarity with the proof of Theorem 3 of Section 6.

> THEOREM 1. If $\underline{g}$ and $\underline{a}$ are recursively enumerable degrees such that $\underline{g} < \underline{a}$, then there exists a recursively enumerable degree $\underline{d}$ such that $\underline{g} \leq \underline{d} < \underline{a}$ and $\underline{d}' = \underline{a}'$.

PROOF. Let f be a recursive function whose range is a set A of degree $\underline{a}$. We define

$$a(s, i) = \begin{cases} 2 & \text{if } (En)(n < s \text{ and } f(n) = i) \\ 1 & \text{otherwise} \end{cases}$$

for all s and i. The function a(s, i) is recursive, and the function $a(i) = \lim_s a(s, i)$ has the same degree as the set A. We define two recursive functions, t(s, e) and h(s, e), for all s and e:

$$t(s, e) = \mu y_{y<s} T_1^1\Big(\prod_{i<y} p_i^{a(s,i)}, e, e, y\Big) \quad ;$$

$$h(0, e) = 0 \quad ;$$

$$h(s + 1, e) = h(s, e) + sg\big(|t(s + 1, e) - t(s, e)|\big) \quad .$$

Let c(e) be the representing function of the predicate

$$(Ey)T_1^1\big(\tilde{a}(y), e, e, y\big)$$

The degree of c is $\underline{a}'$.

LEMMA 1. For each e, $c(e) = 0$ if and only if the set $\{h(s, e) | s \geq 0\}$ is finite.

PROOF. Fix e. Suppose $c(e) = 0$. Let

$$z = \mu y T_1^1\left(\tilde{a}(y), e, e, y\right) .$$

Let v be so large that $v > z$ and $a(v, i) = a(i)$ for all $i < z$. Then $t(s, e) = z$ for all $s \geq v$, and consequently, the set $\{h(s, e) | s \geq 0\}$ is finite.

Now suppose the set in question is finite. Then there must be a w with the property $t(s + 1, e) = t(s, e)$ for all $s \geq w$. Let v be so large that $v > w$ and $a(v, i) = a(i)$ for all $i < t(w, e)$. Then $t(v, e) = t(w, e) \leq w < v$. Let $t(v, e) = t$. Since $t(v, e) < v$, we have

$$T_1^1\left(\prod_{i < t} p_i^{a(v,i)}, e, e, t\right) .$$

Since $a(v, i) = a(i)$ for all $i < t$, we have $T_1^1\left(\tilde{a}(t), e, e, t\right)$, and consequently, $c(e) = 0$.

Let g be an everywhere positive, recursive function whose range is a set G of degree $\underline{g}$. We now proceed exactly as we did in the proof of Theorem 3 of Section 6. We define four recursive functions, $y(s, n, e)$, $m(s, e)$, $r(s, n, e)$ and $d(s, n)$, simultaneously by induction on s. Our definition makes use of the same equations occurring in the proof of Theorem 3 of Section 6; however, it must be remembered that the functions, $a(s, n)$, $h(s, n)$ and $g(s)$, occurring in these equations, are to be taken as defined in the present section rather than as defined in Section 6. We claim that Lemmas 4-9 of Section 6 remain true, despite the new definitions of $a(s, n)$, $h(s, n)$ and $g(s)$. Lemma 1 of the present section is the new proof of Lemma 4 of Section 6. The functions $a(s, n)$ and $g(s)$ of the present section retain all the properties of their counterparts in Section 6 which are needed to prove Lemmas 5-9 of Section 6. In addition, we still have $\underline{a} \not\leq \underline{g}$, a fact needed in the proof of Lemma 8 of Section 6. Let $\underline{c}$ be the degree of c. Lemmas 10 and 11 of Section 6 remain true, since they follow from Lemmas 8 and 9. Thus we have $\underline{a}' = \underline{c} \leq \underline{d}'$ and $\underline{a} \not\leq \underline{d}$. We also have $\underline{g} \leq \underline{d}$ for the same reasons as in Section 6.

We need only show $\underline{d} \leq \underline{a}$. Since $d(n) = 0$ if and only if $(Es)(d(s, n) = 0)$, it will suffice to show the predicate $(Es)(d(s, n) = 0)$ is recursive in the function $a(n)$. Recall that for each $n > 0$,

$$d(s, 5 \cdot 7^n) = 0 \longleftrightarrow n \in G \quad ,$$

and that $d(s, w) = 1$ if w is neither a power of a prime nor of the form $5 \cdot 7^m$ $(m > 0)$. Since G is recursive in $a(n)$, it will suffice to show $(Es)(d(s, p_e^n) = 0)$ is recursive in $a(n)$.

Let $Q(i, m, s, e, n)$ denote the following predicate:

$$i \leq e \ \& \ i \leq m < m(s, i) \ \& \ r(s, m, i) = 1 \ \& \ p_e^n < y(s, m, i) \quad .$$

Then for each e and n,

$$(Es)\Big(d(s, p_e^n) = 0\Big) \longleftrightarrow (Es)\Big(n < h(s, e) \ \& \sim (Ei)(Em)Q(i, m, s, e, n)\Big) \ ,$$

since $d(0, p_e^n) = 1$.

LEMMA 2. The predicate $(Es)\big(n < h(s, e)\big)$ is recursive in the function $a(n)$.

PROOF. We define

$$S = \Big\{ e \mid \text{the set } \{h(s, e) \mid s \geq 0\} \text{ is finite} \Big\} \quad .$$

It follows from Lemma 1 immediately above that the set S is recursively enumerable in the function $a(n)$. For each e, let

$$T_e = \{h(s, e) \mid s \geq 0\} \quad .$$

The sets $T_0, T_1, T_2, \ldots$ are simultaneously recursively enumerable. Fix e and n. We show how to decide whether or not $(Es)\big(h(s, e) > n\big)$. We first recursively enumerate S and T_e simultaneously in a. Eventually either (i) or (ii) will happen:

(i) we will find a member of T_e greater than n;

(ii) we will find that e is a member of S.

If (i) happens, then $(Es)\big(h(s, e) > n\big)$. Suppose (ii) happens. Then, as we saw in the second half of the proof of Lemma 1, there must be a v with the property that $a(v, i) = a(i)$ for all $i < t(v, e)$ and $t(v, e) < v$. We can certainly find such a v with the help of the function $a(n)$. But then $t(s, e) = t(v, e)$ for all $s \geq v$, since $a(s, i) = a(i)$ for all

$s \geq v$ and $t(v, e) < v$. This means $h(s, e) = h(v, e)$ for all $s \geq v$. Then $(Es)\big(h(s, e) > n\big)$ if and only if $h(v, e) > n$.

We introduce the predicate $R(e, n, s)$:

$$(m)_{m>0}(t)\Big[e \leq t \leq n \ \& \ 5 \cdot 7^m < y(s, t, e) \rightarrow d(s - 1, 5 \cdot 7^m) = d(5 \cdot 7^m)\Big]$$

$$\& \ (i)(m)(t)\Big[i < e \leq t \leq n \ \& \ p_i^m < y(s, t, e) \rightarrow d(s - 1, p_i^m) = d(p_i^m)\Big] \ .$$

LEMMA 3. $Q(i, m, s, e, n) \ \& \ R(i, m, s) \rightarrow (t)_{t \geq s} Q(i, m, t, e, n)$.

PROOF. Fix i, m, s, e and n and suppose $Q(i, m, s, e, n)$ and $R(i, m, s)$ hold. We prove $(t)_{t \geq s} Q(i, m, t, e, n)$ by induction on t. Fix $t \geq s$ and suppose $Q(i, m, t, e, n)$ and $R(i, m, t)$ hold. It follows from remark (R3) of Section 6 and Case 1 of the definition of $m(t, i)$ that

$$(n)[i \leq n \leq m \rightarrow y(t, n, i) > 0] \ ,$$

since $i \leq m < m(t, i)$ and $p_e^n < y(t, m, i)$. It follows from $R(i, m, t)$ that

$$(n)[i \leq n \leq r \rightarrow r(t, n, i) = 1] \ ;$$

But then by Lemma 5 of Section 6, we have

$$(n)[i \leq n \leq m \rightarrow y(t, n, i) = y(t + 1, n, i)] \ ,$$

and by Lemma 6 and remark (R1) of Section 6, we have

$$m(t + 1, i) > m \ .$$

Thus $R(i, m, t + 1)$ holds, and consequently, $r(t + 1, m, i) = 1$. But then we have $Q(i, m, t + 1, e, n)$.

LEMMA 4. $(v)(Es)_{s \geq v}(i)(m)[Q(i, m, s, e, n) \rightarrow R(i, m, s)]$.

PROOF. Fix e, n and v. For each $i \leq e$, we define $m(i)$ as we did in Lemma 9 of Section 6. If $i \leq e$ and $i \leq m < m(s, i)$, then $\lim_s y(s, m, i)$ exists and is positive. Let y be so large that

$$(s)(m)\Big[i \leq e \ \& \ i \leq m < m(i) \rightarrow y(s, m, i) \leq y\Big] \ .$$

Let v be so large that

$$(s)(n)\Big[s > v \ \& \ n \leq y \rightarrow d(s-1, n) = d(n)\Big] .$$

By Lemma 7 of Section 6, there is an $s > v$ such that for any unstable $i \leq e$, we have

$$m(s, i) \leq n_j \vee r(s, n_j, i) = 0 \vee y(s, n_j, i) = 0 ,$$

where $e_j = i$; recall that if i is unstable, then $n_j = m(i)$.

Fix i and m and suppose $Q(i, m, s, e, n)$ holds. We show $R(i, m, s)$. Thus we have

$$i \leq e \ \& \ i \leq m < m(s, i) \ \& \ r(s, m, i) = 1 \ \& \ p_e^n < y(s, m, i) .$$

First we suppose i is stable. Then $m < m(s, i) \leq m(i)$, and consequently,

$$(n)(t)\Big[i \leq t \leq m \ \& \ n < y(s, t, i) \rightarrow d(s-1, n) = d(n)\Big] ,$$

since $s > v$. But then $R(i, m, s)$ holds. Now we suppose i is not stable. Let $i = e_j$. Then

$$m(s, i) \leq m(i) \vee r(s, m(i), i) = 0 \vee y\big(s, m(i), i\big) = 0 .$$

If $m(s, i) \leq m(i)$, then $m < m(i)$, and again, $R(i, m, s)$ holds. If $r(s, m(i), i) = 0$, then $m < m(i)$, because $r(s, m, i) = 1$; this last follows from remark (R2) of Section 6. Suppose $y(s, m(i), i) = 0$. Then it follows from remark (R3) of Section 6 that either $m(s, i) \leq m(i)$ or $m(i) = i$. If $m(s, i) \leq m(i)$, then $m < m(i)$, and all is well. Suppose $m(i) = i$. Then $y(s, i, i) = 0$, and it is clear from Case 1 of the definition of $m(s, i)$ that $m(s, i) = i + 1$. But then $m = i$ and $y(s, m, i) = 0$. This last is impossible, since $p_e^n < y(s, m, i)$.

LEMMA 5. $\underline{d} \leq \underline{a}$.

PROOF. It is sufficient to show the predicate $(Es)\big(d(s, p_e^n) = 0\big)$ is recursive in the function $a(n)$. Our argument is informal. We fix e and n and describe procedure P for determining whether or not $(Es)\big(d(s, p_e^n) = 0\big)$. The procedure P will be such that it will readily transform into a system of equations which define d recursively in a. We begin by asking if $(Es)\big(h(s, e) > n\big)$. By Lemma 2, we know the predicate $(Es)\big(h(s, e) > n\big)$ is recursive in the function $a(m)$. If the answer is no, then $\sim (Es)\big(d(s, p_e^n) = 0\big)$. Suppose the answer is yes.

Let

$$v = \mu s\big(h(s, e) > n\big) \quad .$$

Then $(Es)\big(d(s, p_e^n) = 0\big)$ if and only if

$$(Es)_{s \geq v}(i)(m) \sim Q(i, m, s, e, m) \quad .$$

By Lemma 4, we know there is an $s \geq v$ such that

$$(i)(m)\Big[Q(i, m, s, e, m) \rightarrow R(i, m, s)\Big] \quad .$$

Let w be the least such s. Consider what is required for the determination of w. First note that Q is recursive and that i and m are effectively bounded. For a fixed i, m and s, we can decide if $R(i, m, s)$ is true if we know finitely many values of the representing function of G and finitely many values of $d(p_i^n)$ for $i < e$. In the course of determining w, we will learn whether or not $(Ei)(Em)Q(i, m, w, e, n)$. If no such i and m exist, then $d(w, p_e^n) = 0$. Suppose such an i and m do exist. Then by Lemma 3, we have $(s)_{s \geq w}Q(i, m, s, e, n)$. But then we merely have to check all $s < w$ to determine whether or not $(Es)\big(d(s, p_e^n) = 0\big)$. Thus the value of $d(p_e^n)$ can be expressed in terms of the values of $d(p_i^n)$ $(i < e)$ with the help of a predicate of degree $\underline{a} \cup \underline{g} = \underline{a}$.

COROLLARY 1. If $\underline{a}$ and $\underline{b}$ are recursively enumerable degrees such that $\underline{a} < \underline{b}$ and $\underline{a}' < \underline{b}'$, then there is a recursively enumerable degree $\underline{c}$ such that $\underline{a} < \underline{c} < \underline{b}$.[†]

PROOF. By Theorem 1, there is a recursively enumerable degree $\underline{c}$ such that $\underline{a} \leq \underline{c} < \underline{b}$ and $\underline{c}' = \underline{b}'$. Then $\underline{a} < \underline{c}$, since $\underline{a}' < \underline{c}' = \underline{b}'$.

It is immediate from Theorem 2 of Section 6 that the converse of Corollary 1 above is not true; i.e., there are recursively enumerable degrees $\underline{a}$, $\underline{b}$ and $\underline{c}$ such that $\underline{a} < \underline{c} < \underline{b}$ and $\underline{a}' = \underline{b}'$.

† Note added in proof: we now can prove Corollary 1 without the assumption that $\underline{a}' < \underline{b}'$; this result will appear elsewhere.

§8. MINIMAL UPPER BOUNDS FOR SEQUENCES OF DEGREES

In Section 2 we rendered a theorem of Spector [25] to the effect that no infinite, ascending sequence of degrees has a least upper bound. In [25] Spector raised but did not answer the question of whether or not any infinite, ascending sequence has a minimal upper bound. We show below that every countable set of degrees has a minimal upper bound; that is, if D is a countable set of degrees, then there exists a degree $\underline{c}$ greater than or equal to every member of D such that there does not exist a degree $\underline{b}$ greater than or equal to every member of D and less than $\underline{c}$. We also show that the set of all arithmetical degrees has a minimal upper bound which is less than $\underline{0}^{(\omega)}$. Kleene and Post [9] proved that $\underline{0}^{(\omega)}$ is not a minimal upper bound for the set of all arithmetical degrees.

Our arguments in this section are modifications of Spector's construction of a minimal degree.

Let f_0, f_1 and g be functions from the natural numbers into the natural numbers such that f_0 and f_1 are representing functions of sets. We say (f_0, f_1, g) is an admissible triple if the following conditions are met:

(T1) $g(0) = 0 \ \& \ (n)[g(n) < g(n+1)]$;

(T2) $(n)_{n>0}(Et)[g(n) \leq t < g(n+1) \ \& \ f_0(t) \neq f_1(t)]$.

With each admissible triple (f_0, f_1, g) we associate a set $F(f_0, f_1, g)$ of functions as follows:

$$h \in F(f_0, f_1, g) \longleftrightarrow (n)(Ei)_{i<2}(t)[g(n) \leq t < g(n+1) \rightarrow h(t) = f_i(t)] .$$

Observe that $F(f_0, f_1, g)$ is a closed subset of the space F defined in Section 10. If (f_0, f_1, g) and (u_0, u_1, v) are admissible triples, then we say (u_0, u_1, v) is a contraction of (f_0, f_1, g) if

$u_0, u_1 \in F(f_0, f_1, g)$, the range of v is a subset of the range of g, and

$$\mu n\big(u_0(n) \neq u_1(n)\big) > \mu n\big(f_0(n) \neq f_1(n)\big) \quad ;$$

it follows $F(u_0, u_1, v) \subset F(f_0, f_1, g)$.

We say s is a finite initial segment of a function z if s is a partial function whose domain is a finite initial segment of the natural numbers and whose values are given by: $s(n) = z(n)$, when $s(n)$ is defined. The partial function $\{e\}^s$ was defined in Section 1. If s is an initial segment of z, then for each n, either $\{e\}^s(n)$ is undefined or $\{e\}^s(n) = \{e\}^z(n)$.

LEMMA 1. Let (f_0, f_1, g) be an admissible triple. Then for each natural number e, there exists an admissible triple (u_0, u_1, v) with the following properties: (u_0, u_1, v) is a contraction of (f_0, f_1, g); u_0, u_1 and v are recursive in f_0, f_1, g; if $h \in F(u_0, u_1, v)$ then either $\{e\}^h(n)$ is undefined for some n or $\{e\}^h$ is recursive in f_0, f_1, g or h is recursive in $\{e\}^h$, f_0, f_1, g.

PROOF. We take the liberty of writing $s \in^* F(f_0, f_1, g)$ when s is an initial segment of some member of $F(f_0, f_1, g)$ and the domain of s equals $\{t | t < g(n)\}$ for some n.

The definition of (u_0, u_1, v) has three cases.

CASE 1. $(Es)(En)(w)\Big[s \in^* F(f_0, f_1, g)$
$\quad \& \Big(w \in^* F(f_0, f_1, g)$
$\quad \& \ w \text{ extends } s \rightarrow \{e\}^w(n) \text{ not defined}\Big)\Big]$.

Let s and n have the properties assumed in the case hypothesis; let m be such that the domain of s equals $\{t | t < g(m)\}$. We define:

$$m^* = \mu n\Big(n \geq m + 2 \ \& \ (Et)\big(g(n) \leq t < g(n+1) \ \& \ f_0(t) \neq f_1(t)\big)\Big);$$

$$v(0) = 0 \quad ;$$

$$v(t) = g(t + m^* - 1) \quad \text{when} \quad t > 0 \quad ;$$

$$u_i(t) = s(t) \quad \text{when} \quad t < g(m) \quad \text{and} \quad i < 2 \quad ;$$

$$u_i(t) = f_0(t) \quad \text{when} \quad g(m) \leq t < g(m^*) \quad \text{and} \quad i < 2 \quad ;$$

$$u_i(t) = f_i(t) \quad \text{when} \quad t \geq g(m^*) \quad .$$

CASE 2. Case 1 is false; in addition,

$$(Es)(n)(u)(v)\Big[s \in^* F(f_0, f_1, g) \;\&\; \big(u \in^* F(f_0, f_1, g) \;\&\; v \in^* F(f_0, f_1, g) \;\&\; u \text{ extends } s \;\&\; v \text{ extends } s \;\&\; \{e\}^u(n) \text{ is defined} \;\&\; \{e\}^v(n) \text{ is defined} \rightarrow \{e\}^u(n) = \{e\}^v(n)\big)\Big] .$$

Let s have the properties assumed in the case hypothesis. There is a unique m such that the domain of s equals $\{t \mid t < g(m)\}$. We define u_0, u_1 and v as in Case 1.

CASE 3. Both Case 1 and Case 2 are false. We claim:

$$(s)(y)(En)(Eu)(Ew)\Big[s \in^* F(f_0, f_1, g) \;\&\; y \in^* F(f_0, f_1, g) \rightarrow u \in^* F(f_0, f_1, g) \;\&\; w \in^* F(f_0, f_1, g) \;\&\; u \text{ extends } s \;\&\; w \text{ extends } y \;\&\; \{e\}^u(n) \text{ is defined} \;\&\; \{e\}^w(n) \text{ is defined} \;\&\; \{e\}^u(n) \neq \{e\}^w(n) \;\&\; \ell h(u) = \ell h(w)\Big] . \tag{1}$$

To prove (1), suppose $s, y \in^* F(f_0, f_1, g)$. Since Case 2 fails, there is an n, a u and a t such that

$$u, t \in^* F(f_0, f_1, g) \;\&\; u, t \text{ extend } s \;\&\; \{e\}^u(n) \text{ is defined} \;\&\; \{e\}^t(n) \text{ is defined} \;\&\; \{e\}^u(n) \neq \{e\}^t(n).$$

Since Case 1 fails, there is a w such that

$$w \in^* F(f_0, f_1, g) \;\&\; w \text{ extends } y \;\&\; \{e\}^w(n) \text{ is defined.}$$

We can safely assume

$$\ell h(u) = \ell h(t) = \ell h(w) .$$

Now either $\{e\}^u(n)$ or $\{e\}^t(n)$ is unequal to $\{e\}^w(n)$. Suppose $\{e\}^u(n) \neq \{e\}^w(n)$. Then n, u and w have the properties required by (1).

We need (1) to define u_0, u_1 and v simultaneously by induction. Define $m^* - 1 = \mu n(f_0(n) \neq f_1(n))$. Let $v(0) = 0$, $v(1) = g(m^*)$ and $u_i(m) = f_0(m)$ when $m < v(1)$ and $i < 2$. Fix $t > 0$. Suppose: $u_i(m)$ has been defined for all $m < v(t)$ and $i < 2$. For each $i \leq 2^t$, we define a pair (x_i, y_i) of partial functions with finite domains. Let x_0 and y_0 be the partial function whose domain is empty. Fix i so that $0 < i \leq 2^t$ and suppose (x_{i-1}, y_{i-1}) has been defined. We write

$$i = c_0 \cdot 2^0 + c_1 \cdot 2^1 + \ldots + c_t \cdot 2^t \quad ,$$

where each c_j is either 0 or 1. We assume

$$\text{domain of } x_{i-1} = \text{domain of } y_{i-1} = \{m | v(t) \leq m < z\} \quad ,$$

where $z \geq v(t)$. We define two initial segments, s and y:

$$s(m) = y(m) = u_a(m) \quad \text{if} \quad v(j-1) \leq m < v(j) \ \& \ c_{j-1} = a \quad ;$$

$$s(m) = x_{i-1}(m) \quad \text{if} \quad v(t) \leq m < z \quad ;$$

$$y(m) = y_{i-1}(m) \quad \text{if} \quad v(t) \leq m < z \quad .$$

We assume $s, y \in^* F(f_0, f_1, g)$. It follows from (1) that there exist a natural number n and initial segments u and w such that s, y, n, u and w have the properties described in (1). We define

$$x_i(m) = u(m) \quad \text{and} \quad y_i(m) = w(m)$$

for all m such that $v(t) \leq m < z'$, where $\{m | m < z'\}$ is the domain of u and w. The assumptions we made concerning x_{i-1} and y_{i-1} remain true when $i-1$ is replaced by i. Let $r = 2^t$. We define:

$$v(t+1) = v(t) + \text{cardinality of domain of } x_r \quad ;$$

$$u_0(m) = x_r(m) \quad \text{if} \quad v(t) \leq m < v(t+1) \quad ;$$

$$u_1(m) = y_r(m) \quad \text{if} \quad v(t) \leq m < v(t+1) \quad .$$

The assumption we made concerning the values of $u_i(m)$ for $m < v(t)$ and $i < 2$ remain true when t is replaced by $t+1$.

It is easily checked that (u_0, u_1, v) is an admissible triple and is a contraction of (f_0, f_1, g).

It is necessary to examine (1) in order to see that u_0, u_1 and w are recursive in f_0, f_1, g. Let $Q(s, y, n, u, w)$ be what remains after the five initial quantifiers of (1) are removed. Thus we can render (1) as

$$(s)(y)(En)(Eu)(Ev)Q(s, y, n, u, w) \quad .$$

We claim Q is recursive in f_0, f_1, g. If s is a finite initial segment of a function, we can determine whether or not $s \in^* F(f_0, f_1, g)$ by comparing s with sufficiently large initial segments of f_0, f_1 and g. If u is a finite initial segment, we can determine effectively whether or not $\{e\}^u(n)$ is defined for a given n. If follows Q is recursive in

f_0, f_1, g. But then if Case 3 holds, u_0, u_1 and v are recursive in f_0, f_1, g. If Case 1 or Case 2 holds, then it is immediate that u_0, u_1 and v are recursive in f_0, f_1 and g.

Suppose Case 1 holds. Let $h \in F(u_0, u_1, v)$. Let s and n have the properties described in Case 1. Then s is an initial segment of h. It follows $\{e\}^h(n)$ is undefined, since $h \in F(f_0, f_1, g)$.

Suppose Case 2 holds and that $\{e\}^h(m)$ is defined for all m. We show $\{e\}^h$ is recursive in f_0, f_1, g. Let s be an initial segment with the properties required by the hypothesis of Case 2. Then s is an initial segment of h. Fix n. We show how to compute $\{e\}^h(n)$ from f_0, f_1, g. Since Case 1 does not hold, there is an initial segment w such that

$$w \in^* F(f_0, f_1, g) \ \& \ w \text{ extends } s \ \& \ \{e\}^w(n) \text{ is defined.}$$

We can easily find such a w by examining sufficiently large initial segments of f_0, f_1 and g. Let v be an initial segment of h such that

$$\begin{aligned} v \in^* F(f_0, f_1, g) \ &\& \ v \text{ extends } s \\ &\& \ \{e\}^v(n) \text{ is defined} \\ &\& \ \{e\}^v(n) = \{e\}^h(n) \quad . \end{aligned}$$

But then it follows from the hypothesis of Case 2 that

$$\{e\}^w(n) = \{e\}^v(n) = \{e\}^h(n) \quad .$$

Finally, suppose Case 3 holds and that $\{e\}^h(m)$ is defined for all m. We show h is recursive in f_0, f_1, g, $\{e\}^h$. Fix $t > 0$. We indicate how to obtain the values of h(m) for $v(t) \leq m < v(t+1)$ from f_0, f_1, g, $\{e\}^h$ and the values of h(m) for $m < v(t)$. We make use of the fact that (u_0, u_1, v) is an admissible triple, it follows that for each $j < t$, there is just one $a < 2$ such that

$$(2) \qquad (m)\big(v(j) \leq m < v(j+1) \rightarrow h(m) = u_a(m)\big) \quad .$$

For each $j < t$, we define c_j to be the unique $a < 2$ for which (2) holds. Let

$$i = \begin{cases} 2^t & \text{if } (j)(j < t \rightarrow c_j = 0) \\ c_0 \cdot 2^0 + c_1 \cdot 2^1 + \dots + c_{t-1} \cdot 2^{t-1} & \text{otherwise.} \end{cases}$$

We return to Case 3 and consider the definition of (x_1, y_1). It is clear that $s(m) = y(m) = h(m)$ for all $m < v(t)$. For each $b < 2$, let u_b' be the initial segment defined by:

$$u_b'(m) = h(m) \quad \text{if} \quad m < v(t) \quad ;$$
$$u_b'(m) = u_b(m) \quad \text{if} \quad v(t) \leq m < v(t+1) \quad .$$

Then $u_0'(m) = x_1(m)$ and $u_1'(m) = y_1(m)$ for all m in the domain of x_1 and y_1. Let u and w be the initial segments used to define x_1 and y_1. Then u extends s and w extends y, u_0' extends u, and u_1' extends w. In addition, $\{e\}^u(n)$ and $\{e\}^w(n)$ are defined but are not equal; n is the natural number mentioned in the definition of x_1 and y_1. We have

$$\{e\}^{u_0'}(n) = \{e\}^u(n) \neq \{e\}^w(n) = \{e\}^{u_1'}(n) \quad .$$

We can compute u_0' and u_1' from u_0, u_1 and v and the values of $h(m)$ for $m < v(t)$. Since Q is recursive in f_0, f_1, g we can compute n from f_0, f_1, g. Now, either u_0' or u_1' is an initial segment of h, and just one of

$$\{e\}^{u_0'}(n) \quad \text{and} \quad \{e\}^{u_1'}(n)$$

equals $\{e\}^h(n)$. But then we know $h(m)$ when $v(t) \leq m < v(t+1)$, since we know $\{e\}^h(n)$.

The existence of a minimal degree is an easy consequence of Lemma 1. For each e we define an admissible triple as follows. Let $f_i^0(n) = i$ and $g^0(n) = n$ for all n and all $i < 2$. Suppose (f_0^e, f_1^e, g^e) has been defined for some $e \geq 0$. We define $(f_0^{e+1}, f_1^{e+1}, g^{e+1})$ in two stages. Let 0 be the function which is 0 everywhere. Let

$$m(e) = \mu n\left(f_0^e(n) \neq f_1^e(n)\right) \quad . \tag{3}$$

Note that $m(e) < g^e(2)$. Let

$$p = \begin{cases} \{e\}^0\left(m(e)\right) & \text{if} \quad \{e\}^0\left(m(e)\right) \text{ is defined} \\ 0 & \text{otherwise} \quad . \end{cases}$$

We define an admissible triple (f_0, f_1, g):

$$g(0) = g^e(0) = 0 \quad ;$$
$$g(m) = g^e(m+2) \quad \text{if} \quad m > 0 \quad ;$$

$$f_0(m) = f_1(m) = \begin{cases} f_0^e(m) & \text{if} \quad p \neq f_0^e\big(m(e)\big) \ \& \ m < g^e(2) \\ f_1^e(m) & \text{if} \quad p = f_0^e\big(m(e)\big) \ \& \ m < g^e(2) \end{cases} ;$$

$$f_i(m) = f_i^e(m) \quad \text{if} \quad i < 2 \ \& \ m \geq g^e(2) \quad .$$

If $h \in F(f_0, f_1, g)$, then h is not recursive in 0 with Gödel number e, since $h(m(e)) \neq p$. Now we apply Lemma 1 to (f_0, f_1, g) in order to obtain (u_0, u_1, v). Let $f_0^{e+1} = u_0$, $f_1^{e+1} = u_1$ and $g^e = v$.

Note that $m(0) = 0$. Since $(f_0^{e+1}, f_1^{e+1}, g^e)$ is a contraction of (f_0^e, f_1^e, g^e), we have $m(e) < m(e+1)$ for all e. We define a function h by:

(4) $$h(m) = f_0^e(m) \quad \text{if} \quad m(e-1) \leq m < m(e) \quad .$$

Then h is the unique function with the property that $h \in F(f_0^e, f_1^e, g)$ for all g. We claim h has minimal degree. Clearly, h is not recursive, since for each e, $h \in F(f_0^{e+1}, f_1^{e+1}, g)$ and consequently is not recursive in 0 with Gödel number e. It follows from Lemma 1 and the recursiveness of f_0^0, f_1^0 and g^0 that for each $e > 0$, either $\{e\}^h(n)$ is not defined for some n or $\{e\}^h$ is recursive or h is recursive in $\{e\}^h$.

The above construction of a minimal degree is virtually identical with that given by Spector in [25]. It is easily checked that the degree of h is at most $\underline{0}''$.

THEOREM 1. Each countable set of degrees has a minimal upper bound.

PROOF. Let $\{\underline{a}_i \mid i \geq 0\}$ be a sequence of degrees. For each i, let

$$\underline{b}_i = \underline{a}_0 \cup \underline{a}_1 \cup \ldots \cup \underline{a}_i \quad .$$

An upper bound of the $\underline{b}_i$'s is an upper bound of the $\underline{a}_i$'s, and conversely. Thus it is sufficient to find a minimal upper bound for the $\underline{b}_i$'s. Note that for each i,

$$\underline{b}_i \leq \underline{b}_{i+1} \quad .$$

For each i, let b_i be the representing function of a set of degree $\underline{b}_i$. For each e we define an admissible triple (f_0^e, f_1^e, g^e). Let $f_i^0(2n+1) = i$, $f_i^0(2n) = b_0(n)$ and $g^0(n) = 2n$ for all n and all $i < 2$. Fix $e \geq 0$, and suppose (f_0^e, f_1^e, g^e) has been defined. Let

$$k^e(n) = \mu m[g^e(n+1) \leq m < g^e(n+2) \ \& \ f_0^e(m) \neq f_1^e(m)] \quad ;$$

$k^e(n)$ is defined for all n because (f_0^e, f_1^e, g^e) is an admissible triple. We define an admissible triple (f_0, f_1, g). First we recall that the definition of an admissible triple requires that $f_i^e(m) \in \{0, 1\}$ for all m and all $i < 2$. Let

$$f_i(m) = f_i^e(m) \quad \text{if } \underline{g^e}(2n) \leq m < g(2n+1) \ \& \ i < 2 \quad ;$$

$$f_i(m) = f_0^e(m) \quad \text{if } g^e(2n+1) \leq m < g(2n+2) \ \& \ i < 2$$
$$\& \quad f_0(k^e(2n)) = b_{e+1}(n) \quad ;$$

$$f_i(m) = f_1^e(m) \quad \text{if } g^e(2n+1) \leq m < g^e(2n+2) \ \& \ i < 2$$
$$\& \quad f_1(k^e(2n)) = b_{e+1}(n) \quad ;$$

$$g(m) = g^e(2m) \quad .$$

Note that if $h \in F(f_0, f_1, g)$, then $h(k^e(2n)) = b_{e+1}(n)$ for all n. We apply Lemma 1 of the present section to (f_0, f_1, g) to obtain (u_0, u_1, v). Let

$$f_0^{e+1} = u_0, \quad f_1^{e+1} = u_1 \quad \text{and} \quad g^{e+1} = v \quad .$$

In the remarks following Lemma 1, we saw that a contracting sequence of admissible triples has a unique "intersection." Let h be the unique function such that for each $e \geq 0$, we have $h \in F(f_0^e, f_1^e, g^e)$. We prove for each e:

(5) b_e is recursive in h; and f_0^e, f_1^e and g^e are recursive in b_e,

by induction on e. Clearly, (5) is true when $e = 0$. Fix $e \geq 0$ and suppose (5) is true. By Lemma 1, f_0^{e+1}, f_1^{e+1} and g^{e+1} are recursive in f_0, f_1, g. But f_0, f_1 and g are recursive in $f_0^e, f_1^e, g^e, b_{e+1}$. It follows from (5) that f_0^{e+1}, f_1^{e+1} and g^{e+1} are recursive in b_{e+1}, since b_e is recursive in b_{e+1}. Now

$$h(k^e(2n)) = b_{e+1}(n)$$

for all n. Since k^e is recursive in f_0^e, f_1^e, g^e, we have b_{e+1} recursive in h by (5). Thus $\underline{h}$ is an upper bound for the $\underline{b_i}$'s; Lemma 1 also tells us that $\underline{h}$ is a minimal upper bound. For $e > 0$, either $\{e\}^h(n)$ is undefined for some n, or $\{e\}^h$ is recursive in f_0^{e-1}, f_1^{e-1}, g^{e-1} (hence recursive in b_{e-1}) or h is recursive in $\{e\}^h$, f_0^{e-1}, f_1^{e-1}, g^{e-1} (hence recursive in $\{e\}^h$, b_{e-1}).

It is not hard to show that each infinite, ascending sequence of degrees has a continuum of minimal upper bounds. This last follows from the sort of descriptive set-theoretic arguments given at the end of Section 10.

THEOREM 2. The set of all arithmetical degrees has a minimal upper bound which is less than $\underline{0}^{(\omega)}$.

PROOF. Theorem 1 of the present section provides us with a function h such that $\underline{h}$ is a minimal upper bound for the set of all arithmetical degrees. We examine certain details of the proofs of Lemma 1 and Theorem 1 to see that $\underline{h} \leq \underline{0}^{(\omega)}$ if the right choice of b_i's is made. Let 0_0 denote the function which is everywhere 0. For each $n > 0$, let 0_n be the representing function of the predicate

$$(Ey)T_1^1\left(\tilde{0}_{n-1}(y), (e)_0, (e)_1, y\right) .$$

Let 0_ω be a function of two variables defined by: $0_\omega(n, m) = 0_n(m)$. The degree of 0_ω is $\underline{0}^{(\omega)}$.

Let us imagine that the construction contained in the proof of Theorem 1 has been performed with $b_i = 0_i$ for all i. Then the degree of h is a minimal upper bound for the set of all arithmetical degrees, since each arithmetical function is recursive in 0_n for some n. We claim h is recursive in 0_ω. Consider the proof of Theorem 1. Each of the functions, f_0^{e+1}, f_1^{e+1} and g^{e+1}, was defined recursively in f_0^e, f_1^e, g^e, 0_{e+1} with the help of Lemma 1. For each $i < 2$, let z_i be a function such that

$$f_i^{e+1} = \left\{z_i(e)\right\}^{f_0^e, f_1^e, g^e, 0_{e+1}}$$

for all e; similarly, let z_2 be a function such that

$$g^{e+1} = \left\{z_2(e)\right\}^{f_0^e, f_1^e, g^e, 0_{e+1}} .$$

We claim that z_0, z_1 and z_2 can be chosen so that each has degree less than or equal to $\underline{0}^{(\omega)}$. Fix $e \geq 0$. In the proof of Theorem 1, we pass from f_0^e, f_1^e, f_2^e, 0_{e+1} to f_0, f_1, g in a completely effective way; that is, f_0, f_1 and g are defined explicitly in terms of f_0^e, f_1^e, f_2^e and 0_{e+1}. We pass from f_0, f_1, g to f_0^{e+1}, f_1^{e+1}, g^{e+1} with the help of Lemma 1. There are only three cases in the construction of Lemma 1. We can tell which case holds with the help of a predicate of degree at most

$(\underline{f}_0 \cup \underline{f}_1 \cup \underline{g})''$; furthermore, our procedure for deciding which case holds doesn't depend on what the functions f_0, f_1 and g are. Since our procedure is uniform, and since

$$(\underline{f}_0 \cup \underline{f}_1 \cup \underline{g})'' \leq \underline{0}_{e+3} \quad ,$$

it follows z_0, z_1 and z_2 can be assumed to be recursive in 0_ω, since the 0_e's are uniformly recursive in 0_ω. But then for each $i < 2$, there is a function t_i such that

$$f_i^e = \left\{t_i(e)\right\}^{0_\omega}$$

for all e and such that t_i is recursive in 0_ω. The function m was defined in (3);

$$m(e) = \mu n\left(\left\{t_0(e)\right\}^{0_\omega}(n) \neq \left\{t_1(e)\right\}^{0_\omega}(n)\right) \quad .$$

Clearly, m is recursive in 0_ω. But then h, defined in (4), is recursive in 0_ω.

We still must show $\underline{h} < \underline{0}^{(\omega)}$. We define a function d such that 0_n is recursive in d for every n, 0_ω is not recursive in d, and d is recursive in 0_ω. Theorem 1 of Section 2 concerns sets of functions, A, B and C. Let $A = \{0_i | i \geq 0\}$, $B = 0$ and $C = \{0_\omega\}$. Then Theorem 1 of Section 2 provides us with a d such that 0_i is recursive in d for every i, and such that 0_ω is not recursive in d. In order to see that d can be made recursive in 0_ω, we examine stage $s \geq 0$ of the construction of d as described in the proof of Theorem 1. The induction hypothesis at stage s has three clauses which provide a concrete picture of the state of d just before stage s begins. We assume $a_i = 0_i$ for all i. Let d_s be the partial function whose domain consists of all n such that $d(n)$ has been defined prior to stage s and whose values are given by: $d_s(n) = d(n)$. Clauses 1-3 of the induction hypothesis tell us that the domain of d_s, call it D_s, is recursive. They also tell us that the values of d_s are computable from 0_s, since 0_i is recursive in 0_s for all $i < s$. Stage s has four cases; since $B = 0$, Case 2 is not needed. Case 1 describes how to extend d_s to d_{s+1} with the help of 0_s. In Case 3, we have to know whether or not there exists a finite extension of d_s, call it f, and a natural number n such that

$$\{e\}^{o_0,\ldots,o_m,f}(n)$$

is defined but not equal to $o_\omega(n)$. The answer to such a question is provided by o_{s+1}, since $m < s$, the values of d_s are computable from o_s, and D_s is recursive. (We are using the fact that any "one-quantifier question" about o_s is answered by o_{s+1}.) It follows that in Case 3 we extend d_s to d_{s+1} with the help of o_{s+1}. Case 4 is set up to insure, among other things, that d is not recursive in o_ω. Since we wish to avoid such an outcome, we assume $C = 0$ in Case 4. Then we have to know whether or not

$$\{e\}^{o_0,\ldots,o_m}(r_s)$$

is defined. The answer to this question is provided by o_s, since $m < s$. In summation, we observe that at stage s, d_s is extended to d_{s+1} with the aid of o_{s+1}; furthermore, the process of extension is uniform with respect to s. Since the o_s's are uniformly recursive in o_ω, it follows d is recursive in o_ω.

§9. MINIMAL DEGREES

A degree $\underline{d}$ is called minimal if $\underline{0} < \underline{d}$ and if $\underline{0} < \underline{c} < \underline{d}$ for no degree $\underline{c}$. Spector [25] was the first to construct a minimal degree; the one he defined was less than $\underline{0}''$. Shoenfield [23] asked: is there a minimal degree less than $\underline{0}'$? In this section we answer Shoenfield's question affirmatively. We gave a proof in [20], but the proof below is simpler. In the next section we will see that minimal degrees are quite rare: we will prove that the set of all minimal degrees has measure zero. Observe that if $\underline{d}$ is a minimal degree less than $\underline{0}'$, then it follows from Corollary 1 of Section 5 that $\underline{d}$ is not recursively enumerable.

We used the priority method in Sections 4,5,6 and 7 to construct a variety of recursively enumerable degrees; in the present section we use it to construct a degree which is not recursively enumerable. In general, it appears that the priority method is needed whenever it is necessary to exercise a strict economy in the number of quantifiers that occur in the course of a construction. Kleene and Post constructed two predicates of incomparable degrees, each one of which was expressible in both two-quantifier forms. Then Friedberg [1] and Muchnik [13] applied the priority method to the Kleene-Post construction in order to obtain two predicates of incomparable degrees, each one of which was expressible in one-quantifier form. Spector [25] constructed a predicate of minimal degree expressible in both three-quantifier forms. Now we add the priority method to the Spector construction, shake vigorously, and the result is a predicate of minimal degree expressible in both two-quantifier forms.

For each natural number a, let $\tilde{a}$ be such that

$$(i)\left((\tilde{a})_i = (a)_i \dot{-} 1\right) .$$

The recursive predicate, $T_1^1(x, e, b)$, was first defined by Kleene in [8]; we modify its meaning as follows:

$$T_1^1(x, e, b) \leftrightarrow (Ey)(y \leq \ell h(x) \ \& \ T_1^1(x,e,b,y)) \ .$$

We introduce a recursive function:

$$y(x,e,b) = \mu y(y \leq \ell h(x) \ \& \ T_1^1(x,e,b,y)) \ .$$

We will need two recursive predicates defined in [8]:

$$Cpt(a, b) \leftrightarrow (i)_{i < \ell h(a)}[(a)_i = 0 \vee (b)_i = 0 \vee (a)_i = (b)_i] \ ;$$

$$Ext(a, b) \leftrightarrow Seq(a) \ \& \ Seq(b) \ \& \ Cpt(a, b) \ \& \ b \geq a \ .$$

$Ext(a, b)$ says that a and b are sequence numbers with the property that the finite sequence represented by a extends the one represented by b.

Let $\{a_0, a_1, a_2, \ldots\}$ be a set of sequence numbers such that for each i, $Ext(a_{i+1}, a_i)$ and $a_{i+1} > a_i$. Then there is a unique function f with the property that for each i there is an n such that $\bar{f}(n) = a_i$. If A is a set of sequence numbers such that $a_i \in A$ for all i, then we say f is a member of the closure of A, or in symbols, $f \in \bar{A}$. Thus the set of all functions is the closure of the set of all sequence numbers.

Let W be a recursively enumerable set. We say e is a Gödel number of W if W is the range of the partial recursive function

$$U(\mu z \, T_1(e, n, z)) \ ,$$

or in symbols, $W = W_e$. Let 0 be the function which is everywhere zero.

Before we prove Theorem 1 below, let us recapitulate some aspects of Spector's construction of a minimal degree. Spector's construction was equivalent to defining three functions, $u(s)$, $v(i)$ and $z(i)$, with the following properties:

(SP1) $u(s)$ is recursive in $v(i)$, $z(i)$; also, $v(i)$ and $z(i)$ have degrees less than or equal to $\underline{0}''$;

(SP2) the complement of $W_{v(i)}$ equals $W_{z(i)}$.

(SP3) $W_{v(i)}$ is a set of sequence numbers;

(SP4) $(i)(Et)(s)_{s>t}\big(u(s) \in W_{v(i)}\big)$ and $(s)\Big(Ext\big(u(s+1), u(s)\big) \ \& \ u(s+1) > u(s)\Big)$.

(SP5) for each i and each function $f \in \bar{W}_{v(i)}$, $f \neq \{i\}^0$;

(SP6) for each i and each function $f \in \bar{W}_{v(i)}$, either $\{i\}^f$ is not a function or $\{i\}^f$ is a recursive function or f is recursive in $\{i\}^f$.

Let h be the unique function which is a member of the closure of $\{u(s) \mid s \geq 0\}$. It follows immediately from (SP1)-(SP6) that h is a function of minimal degree less than $\underline{0}''$, since $\underline{0}''$ is not minimal. In our construction of a minimal degree we will define two functions $v(s, i)$ and $u(s)$, with the following properties:

(S1) $u(s)$ and $v(s, i)$ have degrees less than or equal to $\underline{0}'$;

(S2) for each i, $\lim_s v(s, i)$ exists and is denoted by $v(i)$;

(S3) $u(s)$ and $v(i)$ have the properties expressed in (SP3)-(SP6).

There are two main differences between Spector's construction and ours: first, Spector requires that $W_{v(i)}$ be recursive (this follows from (SP2)), while we require only that it be recursively enumerable; second, Spector works directly with the function $v(i)$, while we work instead with a sequence, $v(0, i), v(1, i) v(2, i), \ldots,$ of functions which converge to $v(i)$. Spector needs $W_{v(i)}$ recursive in order to prove (SP6); however, as we shall see, recursive enumerability is sufficient. We are forced to sacrifice the recursiveness of $W_{v(i)}$ in order to prove (S1). We work with $v(s, i)$ instead of $v(i)$ for the very same reason. It is generally true of the priority method that one works with a convergent sequence of functions rather than with its limit in order to avoid an unwanted quantifier. This last observation is readily verified by an examination of the proofs of the theorems of Sections 4,5,6 and 7. For example, in the proof of Theorem 1 of Section 5, we work with the function $c(s, n)$ rather than with the function $c(n) = \lim_s c(s, n)$ so that the sets, D_0 and D_1, will be recursively enumerable in B.

THEOREM 1. For each $\underline{b}$ there exists a degree $\underline{d}$ such that $\underline{b} < \underline{d} < \underline{b}'$ and such that $\underline{b} < \underline{c} < \underline{d}$ for no degree $\underline{c}$.

PROOF. First suppose $\underline{b} = \underline{0}$. We define a recursive predicate and a partial recursive function:

$$H(c, t, e, x, m, d, b) \longleftrightarrow (i)_{i<2}\Big[\mathrm{Ext}\big((x)_i, t\big) \ \&\ T_1\big(c, (m)_i, (d)_i\big) \ \&\ (x)_i = U\big((d)_i\big) \ \&\ T_1^1\big((x)_i, e, b\big)\Big] \ \&\ U\big(y((x)_0,e,b)\big) \neq U\big(y((x)_1,e,b)\big) \ ;$$

$$Y(c, t, e) \simeq \mu x\, H\big(c, t, e, (x)_0, (x)_1, (x)_2, (x)_3\big) \ .$$

The predicate H says: t, $(x)_0$ and $(x)_1$ are sequence numbers; both $(x)_0$ and $(x)_1$ extend t and are members of W_c; if f and g are functions such that $(x)_0$ represents an initial segment of f and $(x)_1$ represents an initial segment of g, then $\{e\}^f(b)$ and $\{e\}^g(b)$ are defined but are not equal.

LEMMA 1. If $Y(c, t, e)$ is defined and equal to y, then $\sim \mathrm{Ext}\big((y)_{0,0}, (y)_{0,1}\big)$ and $\sim \mathrm{Ext}\big((y)_{0,1}, (y)_{0,0}\big)$.

PROOF. By the definition of $Y(c, t, e)$, we have

$$T_1^1\big((y)_{0,0}, e, (y)_3\big) \ \&\ T_1^1\big((y)_{0,1}, e, (y)_3\big) \ .$$

It follows from the definition of T_1^1 that if $(y)_{0,0}$ extends $(y)_{0,1}$ or $(y)_{0,1}$ extends $(y)_{0,0}$, then $y((y)_{0,0},e,(y)_3) = y((y)_{0,1},e,(y)_3)$. This last is impossible, since

$$U\big(y((y)_{0,0},e,(y)_3)\big) \neq U\big(y((y)_{0,1},e,(y)_3)\big) \ .$$

We define a recursively enumerable set of sequence numbers denoted by $W(c, t, e)$:

(a) if $\mathrm{Seq}(t)$, then $t \in W(c, t, e)$;

(b) if $u \in W(c, t, e)$ and $Y(c, u, e)$ is defined and equal to y, then $(y)_{0,0}$ and $(y)_{0,1} \in W(c, t, e)$;

(c) $W(c, t, e)$ has no members other than those given by (a) and (b).

Since Y is partial recursive, it is clear that there exists an effective method of obtaining a Gödel number of $W(c, t, e)$, given c, t and e. Let $V(c, t, e)$ be a recursive function with the property that

$$W(c, t, e) = W_{V(c,t,e)}$$

for all c, t and e. Let R(c, t) be a recursive function with the property that

$$W_{R(c,\ t)} = \left\{u \mid Ext(u,\ t)\ \&\ u \in W_c\right\} \cup \{t\}$$

for all c and t. Thus if t is a sequence number, R(c, t) is a Gödel number of the set consisting of all sequence numbers in W_c which extend t and t itself.

We will define five functions, e(s, i), t(s), u(s), v(s, i) and Q(s, i), simultaneously by induction on s. Each of the five will be of degree at most $\underline{0}'$.

Stage s = 0. Let q be a Gödel number of the set of all sequence numbers. We set e(0, i) = i, t(0) = 0, u(0) = 1, v(0,0) = q, v(0, i+1) = $V\big(v(0,\ i),\ u(0),\ i+1\big)$ and Q(0, i) = 1 for all i.

Before we proceed with stage s > 0, we assume

(A) $(Ee)(i)_{i>e}[Q(s-1,\ i) = 1]$.

It is clear (A) is true if s = 1.

Stage s > 0. We define e(s, i) for all i:

$$e(s,\ 0) = \mu j\big(j > 0\ \ \&\ \ Q(s-1,\ j) = 1\big)\ ;$$
$$e(s,\ i+1) = \mu j\big(j > e(s,\ i)\ \ \&\ \ Q(s-1,\ j) = 1\big)\ .$$

It follows from (A) that e(s, i) is well-defined for all i. Note that e(s, i) > 0 for all i. If e is not the Gödel number of a system of equations, then Y(c, t, e) is not defined for any c and t. Since there are infinitely many e which are not Gödel numbers of systems of equations, it follows from (A) that

$$Y\big(v(s-1,\ e(s,\ i) - 1),\ u(s-1),\ e(s,\ i)\big)$$

is not defined for some i; let n(s) be the least such i. Since e(s, n(s)) > 0, we are free to define

$$t(s) = e(s,\ n(s)) - 1\ .$$

The definition of u(s) has two cases:

CASE 1. n(s) = 0. Then t(s) < e(s, 0), and Q(s-1, i) = 0 whenever $0 \leq i \leq t(s)$. Let

$$m(s) = \mu x(i)_{i<2}\Big[Ext\big((x)_i,\ u(s-1)\big)\ \&\ \sim Ext\big((x)_i,\ (x)_{1 \dot{-} i}\big)\Big]\ .$$

Thus we have two sequence numbers, $(m(s))_0$ and $(m(s))_1$, such that each extends $u(s-1)$, but such that neither extends the other. If the partial recursive function $\{s\}^0(i)$ is defined for all $i < m(s)$, then the sequence number $\{\bar{s}\}^0(m(s))$ has length greater than or equal to both $(m(s))_0$ and $(m(s))_1$, but it cannot extend both. This last follows from the fact that if two sequence numbers have a common extension, then one must extend the other. We define

$$u(s) = \begin{cases} (m(s))_0 & \text{if } (Ei)\big(i < m(s) \;\&\; \{s\}^0(i) \text{ is undefined}\big) \\ (m(s))_1 & \text{if } \mathrm{Ext}\big(\{\bar{s}\}^0(m(s)),\, (m(s))_0\big) \\ (m(s))_0 & \text{if otherwise} \end{cases} .$$

CASE 2. $n(s) > 0$. Then

$$Y\big(v(s-1,\, e(s, n(s)-1) - 1),\, u(s-1),\, e(s, n(s) - 1)\big)$$

is defined; let it equal y. Lemma 1 tells us

$$(i)_{i<2}\Big[\mathrm{Ext}\big((y)_{0,i},\, u(s-1)\big) \;\&\; \sim \mathrm{Ext}\big((y)_{0,i},\, (y)_{0,1 \dot{-} i}\big)\Big] .$$

We set $(m(s))_i = (y)_{0,i}$ for all i, and define $u(s)$ as in Case 1.

We define $v(s, i)$ and $Q(s, i)$ for all i:

$$v(s, i) = \begin{cases} v(s-1, i) & \text{if } i \leq t(s) \\ R\big(v(s, i-1),\, u(s)\big) & \text{if } i = t(s) + 1 \\ V\big(v(s, i-1),\, u(s),\, i\big) & \text{if } i > t(s) + 1 \end{cases} ;$$

$$Q(s, i) = \begin{cases} Q(s-1, i) & \text{if } i \leq t(s) \\ 0 & \text{if } i = t(s) + 1 \\ 1 & \text{if } i > t(s) + 1 \end{cases} .$$

We conclude the construction by observing that (A) remains true when s is replaced by $s + 1$.

For each $s > 0$, it follows from the definition of $m(s)$ and $u(s)$ that

$$\mathrm{Ext}\big(u(s),\, u(s-1)\big) \;\&\; u(s) > u(s-1) .$$

Let h be the unique function which is a member of the closure of $\{u(s) \mid s \geq 0\}$. Thus for each s, there is an n such that $u(s) = \bar{h}(n)$.

It is clear that h is not recursive, since for each $s > 0$, the definition of $u(s)$ guarantees that h is not recursive in 0 with Gödel number s.

Before we prove h has minimal degree, we indicate why h has degree at most $\underline{0}'$. At stage $s = 0$ we chose all values in an effective manner. We began stage $s > 0$ by defining $e(s, i)$ $(i \geq 0)$ in an effective manner from $Q(s-1, i)$ $(i \geq 0)$. Then we defined $n(s)$ and $t(s)$ from $e(s, i)$ $(i \geq 0)$ with the help of a predicate of degree $\underline{0}'$. The predicate was needed to determine whether or not the partial recursive function Y was defined for certain arguments. Then $m(s)$ and $u(s)$ were defined from the value of $Q(s-1, t(s))$ with the help of a second predicate of degree $\underline{0}'$. The latter predicate was needed to determine whether or not the partial recursive function $\{s\}^0(i)$ was defined for all $i < m(s)$. Finally, $v(s, i)$ $(i \geq 0)$ and $Q(s, i)$ $(i \geq 0)$ were defined in an effective manner from $v(s-1, i)$ $(i \geq 0)$ and $Q(s-1, i)$ $(i \geq 0)$. In short, the function h has degree at most $\underline{0}'$ because its definition reduces to knowing whether or not certain partial recursive functions are defined for certain arguments.

LEMMA 2. For each i, $\lim_s v(s, i)$ exists.

PROOF. To prove the lemma, it is sufficient to show

$$(i)(Ew)(s)\big(s > w \rightarrow i \leq t(s)\big) \quad ,$$

since $v(s, i) = v(s-1, i)$ whenever $i \leq t(s)$ and $s > 0$. We proceed with an induction on i. Clearly, $0 \leq t(s)$ for all s. Suppose i and w are such that

$$(s)\big(s > w \rightarrow 0 \leq i-1 \leq t(s)\big) \quad ,$$

We wish to show $i \leq t(s)$ for all sufficiently large s. Suppose otherwise; then there is an infinite set M of positive integers such that

$$(s)\big(s \in M \rightarrow i = t(s) + 1\big) \quad .$$

But then

$$(s)\big(s \in M \rightarrow Q(s, i) = 0 \;\&\; Q(s-1, i) = 1\big) \quad ,$$

since $t(s) + 1 = e(s, n(s))$ whenever $s > 0$. The function Q takes only 0 and 1 as values. It follows there must be an infinite set N of positive integers such that

$$(s)\big(s \in N \rightarrow Q(s, i) = 1 \ \& \ Q(s-1, i) = 0\big) \quad .$$

But then

$$(s)\big(s \in N \rightarrow i > t(s) + 1\big) \quad .$$

This last is impossible, since $i \leq t(s) + 1$ whenever $s > w$.

For each i, let

$$v(i) = \lim_s v(s, i) \quad .$$

Our use of the priority method in the present section takes place in Lemma 2 above. From an abstract point of view, our use of the priority method in the construction of a minimal degree is identical with our use of it in Section 4. Lemma 2 of Section 4 says that for each k, the set

$$\{s \mid R_k \text{ is injured at stage } s\}$$

has cardinality at most 2^k. Thus in both Sections 4 and 9, the kth "requirement" is "injured" at most 2^k times. We have already noted that the main construction of Section 5 has the property that we are able to prove each "requirement" is "injured" finitely often, but that we are unable to produce a recursive function f such that the kth requirement is "injured" at most $f(k)$ times. In Theorems 3 and 4 of Section 6 and Theorem 1 of Section 7, we raised the priority method one step higher in its evolution by allowing the possibility of "requirements" that are "injured" infinitely often. In Theorem 3 of Section 6, infinitely many "requirements" are "injured" infinitely often, and the set

$$\{k \mid \text{the } k^{th} \text{ "requirement" is "injured" finitely often}\}$$

is not even recursive.

We now must show that for each i, the function h is a member of the closure of $W_{v(i)}$. This last will follow from Lemma 3.

LEMMA 3. $(s)(i)[u(s) \in W_{v(s,i)} \ \& \ W_{v(s,i+1)} \subseteq W_{v(s,i)}]$.

PROOF. We prove the lemma by induction on s. Clearly,

$$(i)[u(0) \in W_{v(0,i)} \ \& \ W_{v(0,i+1)} \subseteq W_{v(0,i)}] \ .$$

Fix $s > 0$, and suppose

$$(i)[u(s-1) \in W_{v(s-1,i)} \ \& \ W_{v(s-1,i+1)} \subseteq W_{v(s-1,i)}] \ .$$

We now proceed with an induction on i. It is easily seen that $v(w, 0) = q$ for all w. Thus we have

$$u(s) \in W_{v(s,0)} \ \& \ W_{v(s,1)} \subseteq W_{v(s,0)} \ .$$

Fix $i > 0$ and suppose

$$u(s) \in W_{v(s,i-1)} \ \& \ W_{v(s,i)} \subseteq W_{v(s,i-1)} \ .$$

We must show

$$u(s) \in W_{v(s,i)} \ \& \ W_{v(s,i+1)} \subseteq W_{v(s,i)} \ .$$

We introduce two statements, (B) and (C), whose proof we defer briefly:

(B) $n(s) = 0 \rightarrow u(s) \in W_{v(s,t(s))}$;

(C) $n(s) > 0 \rightarrow u(s) \in W_{v(s,e(s,n(s)-1))}$.

We proceed with the induction step of the induction on i. If $i > t(s)$, then the definition of $v(s, i)$ tells us

$$u(s) \in W_{v(s,i)} \ \& \ W_{v(s,i+1)} \subseteq W_{v(s,i)} \ .$$

Suppose $i \leq t(s)$. Let $n(s) = 0$. By (B), $u(s) \in W_{v(s,t(s))}$. But $v(s,t(s)) = v(s-1, t(s))$ and $v(s, i) = v(s-1, i)$, since $i \leq t(s)$. Thus $u(s) \in W_{v(s-1,t(s))}$. But then with the help of the induction hypothesis concerning $s-1$, we obtain

$$u(s) \in W_{v(s-1,t(s))} \subseteq W_{v(s-1,i)} = W_{v(s,i)} \ .$$

Thus $u(s) \in W_{v(s,i)}$. If $i < t(s)$, then $W_{v(s,i+1)} \subseteq W_{v(s,i)}$ is a consequence of the induction hypothesis on $s-1$, since $v(s, i+1) = v(s-1, i+1)$ if $i < t(s)$. If $i = t(s)$, then

$$v(s, i+1) = R(v(s, i), u(s)),$$

and $W_{v(s,i+1)} \subseteq W_{v(s,i)}$, since $u(s) \in W_{v(s,i)}$.

Thus all is well if $n(s) = 0$. Suppose now $n(s) > 0$. We still have $i \leq t(s)$. (C) tells us $u(s) \in W_{v(s,e(s,n(s)-1))}$. Recall that $e(s, n(s)-1) \leq t(s)$. Then $u(s) \in W_{v(s-1,e(s,n(s)-1))}$, since $v(s-1, e(s, n(s)-1)) = v(s, e(s, n(s) - 1))$. If $i < e(s, n(s) - 1)$, then it follows from the induction hypothesis on $s-1$ that

$$u(s) \in W_{v(s-1,e(s,n(s)-1))} \subseteq W_{v(s-1,i)} = W_{v(s,i)} \quad ,$$

and that

$$W_{v(s,i+1)} = W_{v(s-1,i+1)} \subseteq W_{v(s-1,i)} = W_{v(s,i)} \quad .$$

Suppose $e(s, n(s) - 1) \leq i \leq t(s)$. If $i = e(s, n(s) - 1)$, then $u(s) \in W_{v(s,i)}$. Suppose $e(s, n(s) - 1) < i \leq t(s)$. Then $Q(s-1, i) = 0$. Let r be such that $r < s$ and

$$(w)_{r \leq w < s}[Q(w, i) = 0] \ \& \ Q(r-1, i) = 1 \quad ;$$

r is well-defined, since $Q(0,i) = 1$. At stage r, we must have defined:

$$v(r, i) = R\big(v(r, i-1), u(r)\big) \quad .$$

Since $(w)_{r \leq w < s}[Q(w, i) = 0]$, it follows $(w)_{r < w < s}[i \leq t(w)]$. But then we have $v(w, i) = v(r, i)$ whenever $r < w < s$. The same applies to $i - 1$. Thus

$$v(s-1, i) = R\big(v(s-1, i-1), u(r)\big) \quad .$$

Since $i \leq t(s)$, $v(s, i - 1) = v(s - 1, i - 1)$ and consequently

$$v(s-1, i) = R\big(v(s, i-1), u(r)\big) \quad .$$

The induction hypothesis on i tells us $u(s) \in W_{v(s,i-1)}$. It follows from the definition of R that $u(s) \in W_{v(s-1,i)} = W_{v(s,i)}$. Thus we have established $u(s) \in W_{v(s,i)}$ when $e(s, n(s) - 1) \leq i \leq t(s)$. We still must finish showing $W_{v(s,i+1)} \subseteq W_{v(s,i)}$. If $i < t(s)$, then we apply the induction hypothesis on $s - 1$. If $i = t(s)$, then

$$v(s, i+1) = R\big(v(s, i), u(s)\big) \quad ,$$

and $W_{v(s,i+1)} \subseteq W_{v(s,i)}$, since $u(s) \in W_{v(s,i)}$.

It remains only to prove (B) and (C). Let $n(s) = 0$. Then $t(s) < e(s, 0)$ and $Q(s-1, i) = 0$ whenever $0 \leq i \leq t(s)$. For each positive $i \leq t(s)$, let $r(i)$ be such that $r(i) < s$ and

$$(w)_{r(i) \leq w < s}[Q(w, i) = 0] \ \& \ Q(r(i)-1, i) = 1 \quad .$$

By making use of the argument occurring in the previous paragraph, we obtain

$$v(s-1, i) = R\big(v(s-1, i-1), u(r(i))\big)$$

for all positive $i \leq t(s)$. We know $v(s-1, 0) = q$. It follows easily that

$$u(s) \in W_{v(s-1,0)} = W_q,$$

$$u(s) \in W_{v(s-1,1)} = W_{R(v(s-1,0),u(r(1))},$$

$$\ldots \quad \ldots$$

$$u(s) \in W_{v(s-1,t(s))} = W_{R(v(s-1,t(s)-1),\, u(r(t(s))))},$$

since each $r(i)$ is less than s. Recall that $v(s-1, t(s)) = v(s, t(s))$.

To prove (C), we suppose $n(s) > 0$. Let

$$e = e(s, n(s) - 1) \quad .$$

Let r be such that $r < s$ and

$$(w)_{r \leq w < s}[v(w, e) = v(r, e)] \ \& \ [r > 0 \rightarrow v(r-1, e) \neq v(r, e)]$$

We claim $Q(w, e) = 1$ whenever $r \leq w < s$. It follows from the definition of e that $Q(s-1, e) = 1$. Suppose w is such that $r < w < s$ and

$$Q(w-1, e) = 0 \ \& \ Q(w, e) = 1 \quad .$$

Then $e > t(w) + 1$. But then $v(w, e) = V(v(w, e-1), u(w), e)$, and every sequence number in $W_{v(w,e)}$ must be an extension of $u(w)$. This last is impossible, since $r < w$, $v(r, e) = v(w, e)$, and $W_{v(r,e)}$ must contain some member of the form $u(i)$ for some $i \leq r$; clearly, $u(i) < u(w)$. Thus we have $Q(w, e) = 1$ whenever $r \leq w < s$, since $Q(s-1, e) = 1$. If $r = 0$, then $e \leq t(1) = 0$ and $v(r, e) = q = v(s-1, e)$, and $u(s) \in W_{v(s-1,e)} = W_{v(s,e)}$, since $e \leq t(s)$. Suppose then that $r > 0$. Then $v(r-1, e) \neq v(r, e)$, and consequently, $e \geq t(r) + 1$. Since $Q(r, e) = 1$, we can conclude $e > t(r) + 1$. It follows

$$v(r, e) = V\big(v(r, e-1), u(r), e\big) \quad .$$

Since $(w)_{r \leq w < s}[v(w, e) = v(r, e)]$, and since $u(r) \in W_{v(r,e)}$, it must be that $(w)_{r < w < s}[e \leq t(w)]$. But then $(w)_{r \leq w < s}[v(w, e-1) = v(r, e-1)]$. It follows

$$v(s-1, e) = V\big(v(s-1, e-1), u(r), e\big) \quad .$$

Since $n(s) > 0$, $Y\big(v(s-1, e-1), u(s-1), e\big)$ is defined; let it equal y. The induction hypothesis on $s-1$ tells us that $u(s-1) \in W_{v(s-1,e)}$. But then $(y)_{0,i} \in W_{v(s-1,e)}$ for all $i < 2$. Since $u(s) = (y)_{0,i}$ for some $i < 2$, and since $v(s, e) = v(s-1, e)$, we have $u(s) \in W_{v(s,e)}$.

It is now clear that for each i, $h \in \bar{W}_{v(i)}$. Fix i. By Lemma 3,

$$u(s) \in W_{v(s,i)}$$

for all s. But $v(s, i) = v(i)$ for all sufficiently large s. Thus $u(s) \in W_{v(i)}$ for all sufficiently large s, and consequently, $h \in \bar{W}_{v(i)}$.

In the proof of Lemma 2, we showed

$$(i)(Ew)(s)[s > w \rightarrow i \leq t(s)] \quad .$$

It follows immediately that for each i, $\lim_s Q(s, i)$ exists and is equal to either 0 or 1. Lemmas 4 and 6 will suffice to show h has minimal degree.

LEMMA 4. If $\{i\}^h(n)$ is defined for all n, and if $\lim_s Q(s, i) = 0$, then the function $\{i\}^h$ is recursive.

PROOF. Let r be the least w with the property that $Q(s, i) = 0$ for all $s \geq w$. Since $Q(0, i) = 1$, we have $r > 0$. Thus $Q(r-1, i) = 1$, $Q(r, i) = 0$, and consequently,

$$i = t(r) + 1 \quad .$$

Since $Q(s, i) = 0$ for all $s \geq r$, we have $i \leq t(s)$ for all $s > r$. It is clear from the definition of $t(r)$ that

$$Y\big(v(r-1, i-1), u(r-1), i\big)$$

is undefined. Since $i-1 \leq t(s)$ for all $s \geq r$, we have $v(r-1, i-1) =$

$v(i-1)$. Thus $Y\big(v(i-1), u(r-1), i\big)$ is undefined. We saw above (as a consequence of Lemma 3) that $u(s) \in W_{v(i-1)}$ for all sufficiently large s. We need only these last two facts in order to show $\{i\}^h$ is recursive.

We introduce a recursive predicate L:

$$\begin{aligned} L(x, n) \longleftrightarrow \mathrm{Ext}\big((x)_0, u(r-1)\big) & \;\&\; T_1\big(v(i-1), (x)_1, (x)_2\big) \\ & \;\&\; (x)_0 = U\big((x)_2\big) \\ & \;\&\; T_1^1\big((x)_0, i, n, (x)_3\big) \\ & \;\&\; (x)_3 \leq \ell h\big((x)_0\big) \quad . \end{aligned}$$

$L(x, n)$ says: $(x)_0$ is a sequence number which extends $u(r-1)$ and which is a member of $W_{v(i-1)}$; if f is a function such that $(x)_0$ represents an initial segment of f, then $\{i\}^f(n)$ is defined and is equal to $U\big((x)_3\big)$. We claim that for each n, there is an x such that

$$L(x, n) \;\&\; \{i\}^h(n) = U\big((x)_3\big) \quad .$$

Fix n. Since $\{i\}^h(n)$ is defined, we know $(Ey)T_1^1\big(u(s), i, n, y\big)$ & $\{i\}^h(n) = U(y)$ & $y \leq \ell h\big((u(s))\big)$ holds for infinitely many s.[†] Since $u(s) \in W_{v(i-1)}$ for all sufficiently large s, there is no trouble in finding the desired x. To show $\{i\}^h$ recursive, it is enough to show that for every x and n,

$$L(x, n) \rightarrow \{i\}^h(n) = U\big((x)_3\big) \quad .$$

Suppose otherwise. Then there would exist x, y and n such that

$$L(x, n) \;\&\; L(y, n) \;\&\; U\big((x)_3\big) \neq U\big((y)_3\big) \quad .$$

Let $x_0 = x$ and $y_0 = y$. Then we have

$$H\big(v(i-1), u(r-1), i, (z)_0, (z)_1, (z)_2, n\big) \quad ,$$

and consequently, $Y\big(v(i-1), u(r-1), i\big)$ is defined, which it is not.

[†] Recall the remarks made in Section 1 concerning the function Cv.

LEMMA 5. If $u \in W(c, t, i)$ & $w \in W(c, t, i)$ & $w > u$ & $Ext(w, u)$, then $Y(c, u, i)$ is defined and $Ext\big(w, (y(c, u, i))_{0,j}\big)$ for some $j < 2$.

PROOF. Since $u \in W(c, t, i)$, there must exist $u_0, u_1, \ldots, u_n$ and $y(0), y(1), \ldots, y(n)$ such that

$$u_0 = t \;;$$
$$u_k = \big(Y(c, u_{k-1}, i)\big)_{0,y(k)} \quad (0 < k \leq n) \;;$$
$$u_n = u \;.$$

It is of course possible that $u = t$; in that event we take $n = 0$. Similarly, there exist $w_0, w_1, \ldots, w_m$ and $z(0), z(1), \ldots, z(m)$ such that

$$w_0 = t \;;$$
$$w_k = \big(Y(c, w_{k-1}, i)\big)_{0,z(k)} \quad (0 < k \leq m) \;;$$
$$w_m = w \;.$$

Let d be the largest k such that $u_j = w_j$ for all $j \leq k$. Clearly, $d \leq n$. In addition, $d < m$, since $w > u$. For the sake of a reductio ad absurdum, suppose $d < n$. Then $u_{d+1} = \big(Y(c, w_d, i)\big)_{0,y(d+1)}$ and $w_{d+1} = \big(Y(c, w_d, i)\big)_{0,z(d+1)}$. But $y(d+1) \neq z(d+1)$, since $u_{d+1} \neq w_{d+1}$. It follows from Lemma 1 that $\sim Ext(u_{d+1}, w_{d+1})$ and $\sim Ext(w_{d+1}, u_{d+1})$. This means $\sim Ext(w, u_{d+1})$, since $Ext(w, w_{d+1})$. But then $\sim Ext(w, u)$, since $Ext(u, u_{d+1})$. This last is impossible, so we have $d = n$. Thus, $u = u_n = u_d = w_d < w_m = w$. It follows that $Y(c, u, i)$ is defined and that $Ext\big(w, (Y(c, u, i))_{0,z(d+1)}\big)$.

LEMMA 6. If $\{i\}^h(n)$ is defined for all n, and if $\lim_s Q(s, i) = 1$, then the function h is recursive in the function $\{i\}^h$.

PROOF. Since 0 is not the Gödel number of a system of equations, it is clear $i > 0$. Let r be the least w such that

$$(s)\big(s > w \rightarrow i \leq t(s)\big) \;;$$

the existence of r was established in the proof of Lemma 2. Since $i \leq t(s)$ whenever $s > r$, it must be that

$$v(s, i) = v(s-1, i) \ \& \ v(s, i-1) = v(s-1, i-1)$$

whenever $s > r$. But then $v(i) = v(r, i)$ and $v(i-1) = v(r, i-1)$. Since $Q(r, i) = 1$ and $1 + t(r) \leq i$, we have $v(r, i) = V\big(v(r, i-1), u(r), i\big)$. Thus

$$v(i) = V\big(v(i-1), u(r), i\big) \quad .$$

We saw above (as a consequence of Lemma 3) that $u(s) \in W_{v(i)}$ for all sufficiently large s. We need only these last two facts to show h is recursive in $\{i\}^h$.

We introduce the function p:

$$p(0) = u(r) \quad ;$$

$$p(b+1) = \mu x\Big[x \in W_{v(i)} \ \& \ x > p(b) \ \& \ \mathrm{Ext}\big(x, p(b)\big) \ \& \ (Es)\mathrm{Ext}\big(u(s), x\big)\Big] \quad .$$

The fact that p is well-defined is an immediate consequence of the fact that $u(s) \in W_{v(i)}$ for all sufficiently large s. For each b there is an n such that $\bar{h}(n) = p(b)$. In order to show h is recursive in $\{i\}^h$, it will be enough to show p is recursive in $\{i\}^h$.

Fix b. Since $v(i) = V\big(v(i-1), u(r), i\big)$, it is clear that

$$\begin{aligned} p(b) \in W_{V(v(i-1),u(r),i)} \ & \& \ p(b+1) \in W_{V(v(i-1),u(r),i)} \\ & \& \ p(b+1) > p(b) \\ & \& \ \mathrm{Ext}\big(p(b+1), p(b)\big) \quad . \end{aligned}$$

It follows from Lemma 5 that $Y\big(v(i-1), p(b), i\big)$ is defined and that

$$\mathrm{Ext}\big(p(b+1), (Y(v(i-1), p(b), i))_{0,j}\big)$$

for some $j < 2$.

For each b we define:

$$z(b) = Y\big(v(i-1), p(b), i\big) \quad ;$$

$$w(b) = \mu j \ \mathrm{Ext}\big(p(b+1), (z(b))_{0,j}\big) \quad .$$

We now show for all b that

(D) $\quad p(b+1) = (z(b))_{0,w(b)} \quad ;$

(E) $\quad w(b) = \mu j\Big[U\Big(\ell h\big((z(b))_{0,j}\big)\Big) = \{i\}^h\big((z(b))_3\big)\Big] \quad .$

Since Y is partial recursive, it follows from (D) and (E) that p is recursive in $\{i\}^h$. Fix b and let $z = \big(z(b)\big)_{0,w(b)}$. First we prove (D). Note that $w(b) < 2$, and that consequently, $z \in W_{v(i)}$ and $\mathrm{Ext}\big(z,\ p(b)\big)$. It follows from Lemma 1 that $z > p(b)$. We know from the definition of $w(b)$ that $p(b+1)$ extends z, and we know from the definition of $p(b+1)$ that $(Es)\mathrm{Ext}\big(u(s),\ p(b+1)\big)$. Thus $p(b+1) \geq z$ and $(Es)\mathrm{Ext}\big(u(s),\ z\big)$. But then $p(b+1) = z$, and (D) is proved. Now we prove (E). The definition of Y tells us that for each $j < 2$,

$$T_1^1\Big(\big(z(b)\big)_{0,j},\ i,\ \big(z(b)\big)_3\Big) .$$

Since there is an n such that $\bar{h}(n) = p(b+1) = (z(b))_{0,w(b)}$, it follows that

$$\{i\}^h\Big(\big(z(b)\big)_3\Big) = U\Big(\ell h\Big(\big(z(b)\big)_{0,w(b)}\Big)\Big) .$$

The definition of Y also tells us

$$U\Big(\ell h\Big(\big(z(b)\big)_{0,0}\Big)\Big) \neq U\Big(\ell h\Big(\big(z(b)_{0,1}\big)\Big)\Big) .$$

But then (E) is proved.

Let $\underline{d}$ be the degree of the function h. It follows from Lemmas 4 and 6 that $\underline{d}$ is a minimal degree, since Q takes only 0 and 1 as values. We already have seen that $\underline{d} \leq \underline{0}'$. We consider now how to proceed when $\underline{b} > \underline{0}$. Let f be a function of degree $\underline{b}$. We simply retrace the above argument and relativise all functions and predicates, when appropriate, to f. For example, we define $H_f(c,\ t,\ e,\ x,\ m,\ d,\ b)$ as follows:

$$\begin{aligned}(i)_{i<2}\Big[\mathrm{Ext}\big((x)_i,\ t\big)\ &\&\ T_1^1\big(\bar{f}((d)_i),\ c,\ (m)_i\big)\\ &\&\ (x)_i = U\big((d)_i\big)\\ &\&\ T_1^{1,1}\Big(\bar{f}\big(\ell h((x)_i)\big),\ (x)_i,\ e,\ b\Big)\Big]\\ &\&\ U\Big(\ell h\big((x)_0\big)\Big) \neq U\Big(\ell h\big((x)_1\big)\Big) .\end{aligned}$$

In this manner we obtain a function h which is not recursive in f and whose degree is at most $\underline{b}'$. It has the property that for each i, either $\{i\}^{f,h}(n)$ is undefined for some n, or $\{i\}^{f,h}$ is recursive in f, or h is recursive in $\{i\}^{b,h}$, f. Let $\underline{d}$ be the degree of $2^f \cdot 3^h$. Then we have $\mathbf{b} < \mathbf{d} \leq \mathbf{b}'$ and $\mathbf{b} < \mathbf{c} < \mathbf{d}$ for no degree $\mathbf{c}$. Finally, $\mathbf{d} < \mathbf{b}'$ is a consequence of Corollary 1 of Theorem 1 of Section 5.

COROLLARY 1. For each degree $\mathbf{b}$ there is a degree $\mathbf{d}$ such that $\mathbf{b} < \mathbf{d} < \mathbf{b}'$ and such that $\mathbf{d}$ is not recursively enumerable in $\mathbf{b}$.

PROOF. By Theorem 1, there is a $\mathbf{d}$ such that $\mathbf{b} < \mathbf{d} < \mathbf{b}'$ and such that $\mathbf{b} < \mathbf{c} < \mathbf{d}$ for no $\mathbf{c}$. By Corollary 1 to Theorem 1 of Section 5, $\mathbf{d}$ cannot be recursively enumerable in $\mathbf{b}$.

Corollary 1 was first proved by Shoenfield in [23]. His argument is much more direct than the one above. He makes strong use of the fact that the degrees recursively enumerable in $\mathbf{b}$ are simultaneously recursively enumerable in $\mathbf{b}$ in order to obtain a degree $\mathbf{d}$ between $\mathbf{b}$ and $\mathbf{b}'$ such that $\mathbf{d}$ is not less than or equal to any degree recursively enumerable in $\mathbf{b}$.

§10. MEASURE-THEORETIC, CATEGORY AND DESCRIPTIVE SET-THEORETIC ARGUMENTS

Our purpose in this section is two-fold. We present some methodological results bearing on existence arguments for degrees occurring in earlier sections, and we present some results about degrees which require the notion of measure for expression. We show that some of the results of Section 2 can be obtained by either measure-theoretic or category arguments. As an application of descriptive set-theory, we show how the existence of a continuum of minimal degrees can be deduced from Spector's construction of a minimal degree. (Lacombe in an unpublished paper proved the existence of a continuum of minimal degrees by means of a direct construction.) We fulfill our second purpose by showing that the set of all minimal degrees has measure zero. In addition we state a methodological question raised by Spector [25].

Let $T = \{0, 1\}$, and let F be the cartesian product of a countable infinity of copies of T. Thus F is the set of all sequences of 0's and 1's. We will think of F as the set of all representing functions of sets of natural numbers. We topologize F by giving T the discrete topology and F the product topology. (All topological notions and theorems we use can be found in Kelley [6].) Myhill [15] and Lacombe [11] topologize F as a complete, metric space; however, the topology they choose for F is homeomorphic to the product topology we choose for F. Their choice of the complete metric topology seems to be motivated by a desire to use Baire's category theorem for complete, metric spaces. We will use Baire's category theorem for locally compact, regular spaces.

We define a measure for T by specifying that the measure of each one-element subset of T is one-half. We then assign the product measure to F. (All notions and theorems of measure theory we use can be found in Halmos [5].) Thus F has measure 1, and the set of all representing functions of sets containing the element 3 has measure one-half. The measure we have assigned to F is the familiar Lebesgue measure for the closed unit interval. There is an obvious 1-1 measure-preserving map of the closed unit interval into F which is obtained by writing each real number in dyadic notation. Spector [26] used the measure we have chosen for F to prove the existence of two incomparable hyper-degrees.

THEOREM 1. Let A be a non-recursive set. Then the set of all sets in which A is recursive has measure zero.

PROOF. Fix $e \geq 0$. We will show that the set S of all sets in which A is recursive with Gödel number e has measure zero. Then our theorem will follow by the countable additivity of our measure. Let the measure of S be 4m; we suppose $4m > 0$ and show A recursive.

A basic open set of F is specified by two disjoint, finite subsets of natural numbers. Let U and V be two such subsets. Then the set of all sets of natural numbers which contain U but are disjoint from V is a typical basic open set of F, and is denoted by (U, V). The measure of (U, V) is $2^{-(b+c)}$, where b is the cardinality of U and c is the cardinality of V. Each open set of F is the union of countably many basic open sets. It follows that for each open set G and each $\varepsilon > 0$, there exists a finite sequence, $B_1, B_2, \ldots, B_t$, of basic open subsets of G such that $\cup\{B_i | 1 \leq i \leq t\}$ has measure differing from the measure of G by at most ε.

Since the complement of S is measurable, there exists an open set G which contains the complement of S and which is such that the measure of G exceeds that of the complement of S by at most m. There exist basic open subsets, $B_1, B_2, \ldots, B_t$, of G such that the measure of G exceeds the measure of $B = \cup\{B_i | 1 \leq i \leq t\}$ by at most m. Note that B has a finite description; that is, B is completely specified by a finite sequence of ordered pairs of finite sets of natural numbers. Let s be an initial segment of the representing function of a set of natural numbers. By comparing s with B, we can effectively determine whether or not B contains any function which has s as an initial segment; if the answer is yes, then we can effectively find an initial segment s* which extends s and which has the property that any representing function which has s* as an initial segment must be a member of B; finally, the measure of the set of all representing functions which have s* as an initial segment can be effectively determined.

Let g be the representing function of A. We fix n and give an effective method for computing g(n) which does not depend on n. It is clear that we can recursively enumerate all finite initial segments of representing functions. We attack each such initial segment s in the following manner. Let the length of s by y. We think of s as a partial function whose domain is the set of all natural numbers less than y. We check each $w < y$ to see if

$$T_1^1(\tilde{s}(w),\ e,\ n,\ w)$$

is true for any $w < y$. If there is no such w, we are done with s. Suppose there is such a w. Let s' be the unique initial segment of s of length w. If f is any function which has s' as an initial segment, then $\{e\}^f(n)$ is defined and is equal to

$$U\left(\mu w T(\tilde{s}(w),\ e,\ n,\ w)\right) \quad . \tag{1}$$

The set of all representing functions which have s' as an initial segment is a basic open set of F; call it C. We can effectively determine C - B; in fact, C - B will be a finite union of basic open sets. Let D be C - B. The measure of D can be effectively determined. If $f \in D$, then $\{e\}^f(n)$ is defined and is equal to (1). For each i, let T_i be the set of all functions f in F - B such that $\{e\}^f(n)$ is defined and is equal to i. Then D is a subset of just one of the T_i's. Now the measure of $T_{g(n)}$ is at least 3m; if $i \neq g(n)$, then the measure of T_i is at most m. By attacking a sufficiently large number of initial segments in the above manner, we will develop a finite sequence of basic open subsets of $T_{g(n)}$ whose union has measure at least 2m. But as soon as we locate an i with the property that the measure of T_i is at least 2m, we can safely conclude that $g(n) = i$.

For the sake of stating results concisely, we put a measure on the set of all degrees. Let A be a set of degrees. Let A* be the set of all representing functions of sets whose degrees are members of A. We say A is measurable if and only if A* is measurable. If A is measurable, then its measure is the measure of A*.

COROLLARY 1. Let $\underline{d}$ be a degree greater than $\underline{0}$. Then the set of all degrees incomparable with $\underline{d}$ has measure 1.

PROOF. By Theorem 1 the set of all degrees greater than $\underline{d}$ has measure zero. The set of all degrees less than $\underline{d}$ is countable and so has measure zero.

It readily follows from Corollary 1 that there exists an uncountable set of mutually incomparable degrees. In Section 2 we showed there exists a continuum of mutually incomparable degrees. We do not think this last result can be established on purely measure-theoretic grounds. We state Theorem 2 without proof, since its proof does not require any measure-theoretic ideas not occurring in the proof of Theorem 1.

THEOREM 2. Let $\underline{d}$ be a degree which is not recursively enumerable. Then the set of all degrees in which $\underline{d}$ is recursively enumerable has measure zero.

Let F^2 denote the cartesian product of F with itself. We give F^2 the product measure and the product topology.

THEOREM 3. The set of all degrees $\underline{d}$ such that $\underline{d}$ is the union of two incomparable degrees less than $\underline{d}$ has measure 1.

PROOF. Let S_e be the set of all pairs (f, g) of representing functions of sets such that f is recursive in g with Gödel number e. We claim that S_e is a measurable subset of F^2. For each n, let $S_{n,e} = \{(f, g) \mid \{e\}^g(n)$ is defined and equals $f(n)\}$. To say $\{e\}^g(n)$ is defined for a given e and n is to say g has as an initial segment some member of a countable family of initial segments; the countable family is of course determined by n and e. It follows that for each e and n, the set

$$\{g \mid \{e\}^g(n) \text{ is defined}\}$$

is a countable union of basic open subsets of F. But then $S_{n,e}$ is the product of a basic open subset of F and a union of basic open subsets of F. Since $S_e = \cap \{S_{e,n} \mid n \geq 0\}$, we have that S_e is measurable.

Let S be the set of all pairs (f, g) of representing functions such that f is recursive in g or g is recursive in f. It follows from the measurability of S_e for each e that S is measurable. In addition, the measure of S is zero. This last is a consequence of Theorem 1 and Fubini's theorem; it can be shown more directly by noting that each countable subset of F^2 has measure zero and then applying Fubini's theorem in the manner of Spector [25].

It is well-known that there exists a one-one map of F onto F^2 which is measure-preserving; that is, there exists a one-one map z of F onto F^2 such that for each subset A of F, A is a measurable subset of F with measure m if and only if z(A) is a measurable subset of F^2 with measure m. The map z may be defined as follows. For each function f, let the functions u_f and v_f be defined by $u_f(n) = f(2n)$ and $v_f(n) = f(2n+1)$. Then

$$z(f) = (u_f, v_f) \quad .$$

It is clear that z is one-one and onto. To see that z is measure-preserving, it is sufficient to check the action of z on basic open subsets of F; z maps each basic open subset of F onto a basic open subset of F^2 which is representable as the product of two basic open subsets of F. Since we have given both F and F^2 the product measure, it follows that z is measure-preserving.

Let T be the set of all $f \in F$ such that u_f is recursive in v_f or v_f is recursive in u_f. Then z(T) equals S. Since S has measure zero, it follows that T is measurable and has measure zero. But then our theorem is proved, since the degree of f is the union of the degrees of u_f and v_f.

COROLLARY 1. The set of all minimal degrees has measure zero.

Using the ideas of the proof of Theorem 2, we can also show that the set of degrees $\underline{d}$ such that $\underline{d}$ has only finitely many predecessors has measure zero.

We now turn to category arguments. We use Baire's category theorem in the following form: if X is a locally compact, regular topological space, then the intersection of countably many dense open subsets of X is itself dense in X. By Tychonoff's theorem (or König's Lemma), F is compact, hence locally compact. F is regular because each basic open subset of F is also closed. In [15] Myhill uses Baire's category theorem to obtain an uncountable set of mutually incomparable degrees.

THEOREM 4. Let f be a non-recursive member of F. Then the set of all representing functions g such that g is not recursive in f and f is not recursive in g is dense in F.

PROOF. Let K be the set of all functions g such that g is not recursive in f and f is not recursive in g. Our theorem is an immediate consequence of Baire's category theorem if K contains the intersection of countably many dense open sets. For each e, let

$$P_e = \{g \mid (En)(\{e\}^f(n) \text{ is undefined or unequal to } g(n)\} \ ;$$
$$Q_e = \{g \mid (En)(\{e\}^g(n) \text{ is undefined or unequal to } f(n)\} \ .$$

Then K is the intersection of all the P_e's and Q_e's. In the light of the remarks made at the beginning of the proof of Theorem 3, it is clear that each P_e is open and dense. For each initial segment s of a representing function; we define a non-empty basic open set $R_{e,s}$ as follows:

CASE 1. There exists a natural number n such that for each representing function g which has s as an initial segment, $\{e\}^g(n)$ is undefined. We define $R_{e,s} = \{g \mid g$ has s as an initial segment$\}$.

CASE 2. Case 1 is false, and there exists a natural number n and a finite initial segment t of a representing function such that t extends s and such that for each representing function g which has t as an initial segment, $\{e\}^g(n)$ is defined but not equal to f(n). We define $R_{e,s} = \{g|g$ has t as an initial segment$\}$.

CASE 3. Both Case 1 and Case 2 are false. But then for each natural number n, there is a finite initial segment t such that $\{e\}^t(n)$ is defined and t extends s; in addition, for any such t, $\{e\}^t(n) = f(n)$. But then it follows (as in Case 2.2 of Theorem 1 of Section 2) that f is recursive, which is false.

If B is a basic open set, then $R_{e,s} \subseteq B$ for some s. It follows $\cup \{R_{e,s}|s$ ranges over all finite, initial segments of representing functions$\}$ is a dense, open subset of Q_e.

Let us compare Corollary 1 of Theorem 1 with Theorem 4. Each of these propositions expresses the following fact: if f is a non-recursive function, then the set S of all functions incomparable with f is non-empty. The measure-theoretic proposition says S has the same measure as F, while the category-theoretic proposition says S is a dense subset of F. It seems to us that the measure-theoretic statement is stronger than the category-theoretic statement, because F has dense subsets of measure zero and any subset of F of measure 1 is dense in F. Of course, the measure-theoretic argument has the defect that it is more complicated than the category argument. Corollary 1 of Theorem 3, which says that the set M of all functions whose degrees are minimal has measure zero, does not seem to have a category-theoretic counterpart of equal strength. It may or may not be true that M is a meager (first category) subset of F; however, it is certainly true that F has meager subsets of measure 1.

The category-theoretic view does shed considerable light on what happened in Sections 2 and 3. It does not seem to have much bearing on the more difficult arguments of Sections 5-9. We can, as promised in Section 4, construe Theorem 1 of Section 4 as an effective version of the priority method. Theorem 1 of Section 4 is an abstract rendering of the combinatorial content of the Friedberg-Muchnik solution of Post's problem. We can now restate the definitions of Section 4 as follows. A requirement is an open subset of F. A member of F meets requirement R if it is a member of R. A function t simultaneously enumerates open sets R_1, R_2, R_3, ... in the following manner: for each s, $t(s)$ is a basic open subset of R_k, where $k = (t(s))_2$. Since each open subset of F is a countable union of basic open subsets, it follows that any sequence of open subsets of F can be simultaneously enumerated by some function t. Of course it may not be possible to choose t recursive. Let t enumerate the open sets $R_1, R_2, R_3, \ldots$; Baire's category theorem says that the dense members of the sequence $R_1, R_2, R_3, \ldots$ have a non-empty intersection, and that this intersection has a member which is recursively enumerable in t.

The priority method is needed in the proof of Theorem 1 of Section 4 because we do not know which members of the sequence $R_1, R_2, R_3, \ldots$ are t-dense; more precisely, there does not, in general, exist a set D such that R_k is t-dense if and only if $n \in D$ and such that D is recursive in t. The solution to Post's problem given in Section 4 consists of the definition of a suitable recursive function t which enumerates requirements; it can be shown there is no effective method for determining whether a given requirement enumerated by t is t-dense. This last fact was mentioned in considerably different language by Friedberg in a footnote of [1].

In an unpublished paper, Lacombe constructed a continuum of functions of minimal degree by modifying Spector's construction of a function of minimal degree. Let M be the set of all functions of minimal degree. We show M has cardinality of the continuum by a descriptive set-theoretic argument. (The definitions and theorems we use can be found in Sierpinski [24].) Let $r_1, r_2, r_3, \ldots$ be a listing of all the recursive functions in F. For each e, f, and n, we define the following subsets of F:

$$M^1_{e,n} = \{g \mid \{e\}^g(n) \text{ is defined}\} \;;$$

$$M^2_{e,f,n} = \{g \mid \{e\}^g(n) \text{ is defined and equal to } r_f(n)\} \;;$$

$$M^3_{e,f,n} = \{g \mid \{f\}^{\{e\}^g}(n) \text{ is defined and equal to } g(n)\} \;.$$

Each of the above sets is open, since each is the countable union of the basic open sets. For each e, we define four subsets of F:

$M_e^1 = \{g \mid \{e\}^g$ is not a function$\}$;

$M_e^2 = \{g \mid \{e\}^g$ is a recursive function$\}$;

$M_e^3 = \{g \mid \{e\}^g$ is a function and g is recursive in $\{e\}^g\}$;

$M_e = M_e^1 \cup M_e^2 \cup M_e^3$.

Each of these four sets is Borel, since:

$$M_e^1 = \bigcup_{n \geq 0} (F - M_{e,n}^1) \quad ;$$

$$M_e^2 = \bigcup_{f \geq 0} \bigcap_{n \geq 0} M_{e,f,n}^2 \quad ;$$

$$M_e^3 = \bigcup_{f \geq 0} \bigcap_{n \geq 0} M_{e,f,n}^3 \quad .$$

But then M is Borel, since $M = \bigcap \{M_e \mid e \geq 0\}$. (For our purposes it is sufficient that M be analytic.) Now F is regular, compact, T_1, and has a countable base. It follows that F is a complete, separable metric space, and that consequently, any analytic subset of F is either countable or else has cardinality equal to that of the continuum. We need only show M is uncountable in order to conclude M has cardinality of the continuum. Let $m_0, m_1, m_2, \ldots$ be a sequence of functions of minimal degree. We indicate how to construct a function h of minimal degree which is not a member of the given sequence. Let 0 be the function defined by $0(n) = 0$ for all n. In Section 8 we defined a function h of minimal degree; our construction contained a step in which we made sure h was not recursive in 0 with Gödel number e. This step requires only a trivial modification in order to provide us with a guarantee that h is not recursive in m_i with Gödel number e.

In [25], Spector raised a methodological question connected with his construction of a minimal degree. Let S be a set of natural numbers, and let σ be a partial function which takes only 0 and 1 as values and which is defined on S. Let $F(S, \sigma)$ denote the set of all representing functions g which agree with σ on S: $g(n) = \sigma(n)$ for all $n \in S$. (Note that $F(S, \sigma)$ is a countable intersection of basic open sets of F, is closed, and has measure zero if S is infinite.) Spector [25] asked: given a number e does there exist a set S with infinite complement and a partial function σ defined on S and taking only 0 and 1 as values such that

$$g \in F(s, \sigma) \text{ and } [f \text{ is recursive in } g \text{ with Gödel number } e]$$
$$\rightarrow [f \text{ is recursive}] \text{ or } [g \text{ is recursive in } f] ,$$

where f ranges over all representing functions. Spector then goes on to say that a positive answer to his question would greatly simplify his construction of a function of minimal degree. The answer to Spector's question is unknown. We conjecture that the answer is no.

We conclude this section with a remark on category arguments. Baire's category theorem is proved by means of an ordinary induction; that is, a nested sequence of closed sets is defined by an induction on the natural numbers. It is not surprising, then, that many existence proofs which make use of an ordinary induction can be rephrased in such a manner that they become consequences of Baire's category theorem. (For example, it is possible to prove the existence of a continuous, nowhere differentiable function by means of a category argument.) Theorem 4 of the present section is little more than a rephrasing of some of the details of Theorem 1 of Section 2. It seems to us that category arguments provide an elegant way of presenting elementary results about degrees, but that they are useless when it comes to discovering or presenting difficult results about degrees. As an example of such a difficult result we give Corollary 1 of Theorem 3 of Section 6. We wish to be included among the first to point out that most of the difficult results of this monograph are surrounded by a wilderness of almost incomprehensible equations; however, we do not believe this wilderness can be tamed by means of category arguments alone.

§11. INITIAL SEGMENTS OF DEGREES

Let P be a partially ordered set. A subset Q of P is said to be an initial segment of P if

$$p \in P \;\&\; q \in Q \;\&\; p < q \rightarrow p \in Q \quad .$$

In this section we study the following question: what are the possible finite segments of the upper semi-lattice of degrees less than or equal to $\underline{0}''$. We have already obtained one result in this direction; in Section 9, we showed there exists a minimal degree less than $\underline{0}'$. In this section we obtain two further results: (1) there exists two incomparable degrees, $\underline{b}$ and $\underline{c}$, each less than $\underline{0}''$, such that the only degrees less than $\underline{b} \cup \underline{c}$ are $\underline{b}$, $\underline{c}$ and $\underline{0}$; (2) there exist non-zero degrees $\underline{c}$ and $\underline{d}$ such that the only degrees less than $\underline{d}$ are $\underline{c}$ and $\underline{0}$. The first result answers in the affirmative a question suggested by a result of Titgemeyer [27]; the second answers in the affirmative a question raised by Spector [25] and studied by Shoenfield [23]. The second result was first proved by Titgemeyer [27]. We assume absolute mastery of Section 8.

THEOREM 1. There exist two incomparable degrees, $\underline{b}$ and $\underline{c}$, each less than $\underline{0}''$, such that the only degrees less than $\underline{b} \cup \underline{c}$ are $\underline{b}$, $\underline{c}$ and $\underline{0}$.

PROOF. We make use of the notions of Section 8. Let f_0, f_1 and g be functions from the natural numbers into the natural numbers such that f_0 and f_1 are representing functions of sets. We use the prefix $(E \infty n)$ to mean there exist infinitely many n. We say (f_0, f_1, g) is a special triple if:

(S1) $g(0) = 0 \;\&\; (n)[g(n) < g(n+1)]$;

(S2) $(n)_{n>0}(Et)[g(n) \leq t < g(n+1) \;\&\; f_0(t) \neq f_1(t)]$;

(S3) $(E \infty n)(t)[g(n) \leq 2t < g(n+1) \rightarrow f_0(2t) = f_1(2t)]$;

(S4) $(E \infty n)(t)[g(n+1) \leq 2t + 1 < g(n+2) \rightarrow f_0(2t+1) = f_1(2t+1)]$.

We define contraction of special triples as in Section 8. For each function h we define $Ev(h)(n) = h(2n)$ and $Od(h)(n) = h(2n+1)$ for all n.

LEMMA 1. Let (f_0, f_1, g) be a special triple. Then for each natural number e, there exists a special triple (u_0, u_1, v) with the following properties: (u_0, u_1, v) is a contraction of (f_0, f_1, g); u_0, u_1 and v are recursive in f_0, f_1, g; if $h \in F(u_0, u_1, v)$, then either $\{e\}^h(n)$ is undefined for some n or $\{e\}^h$ is recursive in $f_0, f_1, g, Ev(h)$ or $\{e\}^h$ is recursive in $f_0, f_1, g, Od(h)$ or h is recursive in $\{e\}^h, f_0, f_1, g$.

PROOF. We take the liberty of writing $s \in^* F(f_0, f_1, g)$ when s is an initial segment of some member of $F(f_0, f_1, g)$ and the domain of s equals $\{t \mid t < g(n)\}$ for some n.

The definition of (u_0, u_1, v) has four cases.

CASE 1. $(Es)(En)(w)[s \in^* F(f_0, f_1, g)$ & $(w \in^* F(f_0, f_1, g)$ & w extends $s \rightarrow \{e\}^w(n)$ not defined$)]$.

We define (u_0, u_1, v) as in Case 1 of Lemma 1 of Section 8.

CASE 2. Case 1 is false; in addition,

$$(Es)(n)(u)(v)\Big[s \in^* F(f_0, f_1, g) \;\&\; \big(u \in^* F(f_0, f_1, g) \;\&\; v \in^* F(f_0, f_1, g) \;\&\; u \text{ extends } s \;\&\; v \text{ extends } s \;\&\; (t)(2t \geq \ell h(s) \rightarrow u(2t) = v(2t)) \;\&\; \{e\}^u(n) \text{ is defined} \;\&\; \{e\}^v(n) \text{ is defined} \rightarrow \{e\}^u(n) = \{e\}^v(n)\big)\Big] .$$

Let s have the properties assumed in the case hypothesis. Then proceed as in Case 1.

CASE 3. The same as Case 2 save that "2t" is replaced by "2t+1."

CASE 4. Cases 1, 2, and 3 are all false. We claim both (1) and (2) hold:

(1) $$(s)(y)(En)(Eu)(Ew)\Big[s \in^* F(f_0, f_1, g) \;\&\; y \in^* F(f_0, f_1, g) \;\&\; \ell h(s) = \ell h(y) \rightarrow \big(u \in^* F(f_0, f_0, g) \;\&\; w \in^* F(f_0, f_1, g) \;\&\; \ell h(u) = \ell h(w) \;\&\; (t)(\ell h(s) \leq 2t < \ell h(u) \rightarrow u(2t) = w(2t)$$

$\& \{e\}^u(n)$ is defined $\& \{e\}^w(n)$ is defined

$\& \{e\}^u(n) \neq \{e\}^w(n)\big)\Big]$;

(2) In (1), replace "2t" by "2t+1."

The falsity of Cases 1 and 2 entail (1); the argument is almost identical with the one occurring in Case 3 of Lemma 1 of Section 8. Similarly, the falsity of Cases 1 and 3 entail (2).

We need (1) and (2) to define u_0, u_1 and v simultaneously by induction. Let $m^* = \mu n(f_0(n) \neq f_1(n))$. Let $v(0) = 0$, $v(1) = g(m^*+1)$ and $u_i(m) = f_0(m)$ when $m < v(1)$ and $i < 2$. Fix $t > 0$ and suppose t is even. Suppose: $v(m)$ has been defined for all $m \leq t$; $v(0) < v(1) < \ldots < v(t)$; and $u_i(n)$ has been defined for all $n < v(t)$ and $i < 2$. For each $i \leq 2^t$, we define a pair (x_i, y_i) of partial functions with finite domains. Let x_0 and y_0 be the partial function whose domain is empty. Fix i so that $0 < i \leq 2^t$ and suppose (x_{i-1}, y_{i-1}) has been defined. We write

$$i = c_0 \cdot 2^0 + c_1 \cdot 2^1 + \ldots + c_t \cdot 2^t ,$$

where each c_j is either 0 or 1. We assume

$$\text{domain of } x_{i-1} = \text{domain of } y_{i-1} = \{m | v(t) \leq m < z\} ,$$

where $z \geq v(t)$. We define two initial segments, s and y:

$$s(m) = y(m) = u_a(m) \text{ if } v(j-1) \leq m < v(j) \;\&\; c_{j-1} = a ;$$

$$s(m) = x_{i-1}(m) \text{ if } v(t) \leq m < z$$

$$y(m) = y_{i-1}(m) \text{ if } v(t) \leq m < z .$$

We assume $s, y \in^* F(f_0, f_1, g)$. It follows from (1) that there exists a natural number n and initial segments u and w such that s, y, n, u and w have the properties described in (1). We proceed to define x_i, y_i, n, u_0, u_1, and v as in Case 3 of Lemma 1 of Section 8. We assumed above t is even. If t is odd, we proceed as above save that (1) is replaced by (2).

It is easily checked that (u_0, u_1, v) is a special triple and is a contraction of (f_0, f_1, g). It is necessary to examine (1) and (2) to see that u_0, u_1 and v are recursive in f_0, f_1, g. The reasoning again is identical with that occurring in Lemma 1 of Section 8.

Now we consider the significance of the above four cases.

Suppose Case 1 holds. Then, as in Lemma 1 of Section 8, for each $h \in F(u_0, u_1, v)$, we have $\{e\}^h(n)$ undefined for some n.

Suppose Case 2 holds. Let s be an initial segment with the properties required by Case 2. Let $h \in F(u_0, u_1, v)$ and suppose $\{e\}^h(m)$ is defined for all m. Now s must be an initial segment of h. Fix n. We show how to compute $\{e\}^h(n)$ from $f_0, f_1, g, Ev(h)$. Since $\{e\}^h(n)$ is defined and since $h \in F(f_0, f_1, g)$, there must be an initial segment w such that

$$w \in^* F(f_0, f_1, g) \ \& \ w \text{ extends } s$$
$$\& \ \{e\}^w(n) \text{ is defined } \& \ (t)(w(2t)) = Ev(h)(2t)).$$

We can easily find such a w by examining sufficiently large initial segments of f_0, f_1, g and $Ev(h)$. Let v be an initial segment of h such that

$$v \in^* F(f_0, f_1, g) \ \& \ v \text{ extends } s$$
$$\& \ \{e\}^v(n) \text{ is defined } \& \ \{e\}^v(n) = \{e\}^h(n).$$

But then it follows from the hypothesis of Case 2 that

$$\{e\}^w(n) = \{e\}^v(n) = \{e\}^h(n) \quad .$$

If Case 3 holds and $\{e\}^h(m)$ is defined for all m, then $\{e\}^h$ is recursive in $f_0, f_1, g, Od(h)$ by the same argument as the one immediately above.

If Case 4 holds and $\{e\}^h(m)$ is defined for all m, then h is recursive in $f_0, f_1, g, \{e\}^h$ by the same argument as in Lemma 1 of Section 8.

Theorem 1 of the present section is an easy consequence of Lemma 1. For each e we define a special triple of recursive functions as follows. Let $f_i^0 = i$ and $g^0(n) = n$ for all n and all $i < 2$. Suppose (f_0^e, f_1^e, g^e) had been defined for some $e \geq 0$. We define $(f_0^{e+1}, f_1^{e+1}, g^{e+1})$ in two stages. We assume $e + 1$ is odd. Let

$$(3) \qquad m(e) = \mu n(Em)\Big[g^e(m+1) \leq n < g^e(m+2) \ \& \ (t)\big(g^e(m+1) \leq 2t < g(m+2) \rightarrow f_0^e(2t) = f_1^e(2t)\big) \ \& \ f_0^e(n) \neq f_1^e(n)\Big] \quad ;$$

m(e) exists by clause (S3) of the definition of special triple. Let

$$s \simeq \mu t(Ew)\Big[w \in^* F(f_0^e, f_1^e, g^e) \ \&\ t = Ev(w) \ \&\ \{e\}^t\big((m(e)-1)/2\big) \text{ is defined}\Big] \ ;$$

$$v \simeq \mu w\big(w \in^* F(f_0^e, f_1^e, g^e) \ \&\ s = Ev(w)\big) \ ;$$

$$k \simeq \mu t\big(t \text{ extends } v \ \&\ t(m(e)) \neq \{e\}^s\big((m(e)-1)/2\big) \ \&\ t \in^* F(f_0^e, f_1^e, g^e)\big) \ .$$

Let $w = k$ if k exists; otherwise let k be the partial function whose domain is $\{t \mid t < g^e(1)\}$ and whose values are given by $w(m) = f_0^e(m)$. Then if h is a function such that w is an initial segment of h, we have $Od(h)$ is not recursive in $Ev(h)$ with Gödel number e. Note that if s and v exist, then k exists by definition of $m(e)$. If $e+1$ is even rather than odd, we proceed as above with "even" and "odd" interchanged. We define a special triple (f_0, f_1, g):

$$g(0) = g^e(0) = 0 \ ;$$

$$g(m) = g^e(m+r-1), \quad \text{where } \{t \mid t < g^e(r)\} \text{ is domain of } w \text{ and } m > 0 \ ;$$

$$f_0(m) = f_1(m) = w(m) \quad \text{if } m < g(1) \ ;$$

$$f_i(m) = f_i^e(m) \quad \text{if } i < 2 \ \&\ m \geq g^e(1) \ .$$

Now we apply Lemma 1 of the present section to (f_0, f_1, g) to obtain (u_0, u_1, v). Let $f_0^{e+1} = u_0$, $f_1^{e+1} = u_1$ and $g^{e+1} = v$.

We define h as in (4) of Section 8. Then h is the unique function such that $h \in F(f_0^e, f_1^e, g^e)$, and $Od(h)$ and $Ev(h)$ have incomparable degrees. Let $\underline{b}$ be the degree of $Od(h)$, and let $\underline{c}$ be the degree of $Ev(h)$. Then $\underline{b} \cup \underline{c}$ is the degree of h. It follows from Lemma 1 of the present section that any degree less than $\underline{b} \cup \underline{c}$ is either less than or equal to $\underline{b}$ or less than or equal to $\underline{c}$. Now we did not bother to make $\underline{b}$ and $\underline{c}$ minimal, but this is easy to accomplish with the help of Lemma 1 of Section 8. We observe that $\underline{b} \cup \underline{c}$ can be forced to have degree less than $\underline{0}''$, because we can tell which case holds in Lemma 1 of Section 8 or Lemma 1 of the present section with the help of a certain 2-quantifier form, and because the definition of w entering into the definition of $(f_0^{e+1}, f_1^{e+1}, g^{e+1})$ requires only a 1-quantifier form.

COROLLARY 1. There exists a degree $\underline{d}$ such that $\underline{d}$ is the least upper bound of the set of all degrees less than itself and such that $\underline{d}$ is not recursively enumerable in any degree less than itself.

PROOF. Let $\underline{d} = \underline{b} \cup \underline{c}$, where $\underline{b}$ and $\underline{c}$ are provided by Theorem 2 above. Then $\underline{d}$ has only three predecessors, $\underline{b}$, $\underline{c}$ and $\underline{0}$. It follows from Corollary 4 to Theorem 1 of Section 5 that $\underline{d}$ is not recursively enumerable in any degree less than itself.

Corollary 1 says that the converse of Corollary 3 to Theorem 1 of Section 5 is false.

THEOREM 2. (Titgemeyer [27]) There exist degrees $\underline{b}$ and $\underline{d}$, each less than $\underline{0}''$ and greater than $\underline{0}$, such that the only degrees less than $\underline{d}$ are $\underline{b}$ and $\underline{0}$.

We proceed as in Theorem 1 with one major change. We drop Case 2 of Lemma 1. Thanks to Case 3, we still can show $\underline{c} \nleq \underline{b}$. We are unable to show $\underline{b} \nleq \underline{c}$ since Case 2 is absent. Cases 1, 2, and 4 retain their former significance. Thus any degree less than $\underline{b} \cup \underline{c}$ is less than or equal to $\underline{b}$ by Case 3. Let $\underline{d} = \underline{b} \cup \underline{c}$. Then $\underline{b} < \underline{d}$. We must make one minor change: $\underline{c}$ is no longer minimal; if we attempt to make $\underline{c}$ minimal, we will fail in our attempt to make $\underline{b}$ minimal. In order to make $\underline{b}$ minimal, we must require

$$\text{(S5)} \quad (E \infty n)(Et)\left(g^e(n) \leq 2t < g^e(n+1) \ \&\ f_0^e(2t) \neq f_1^e(2t)\right) .$$

for all e; in other words there must be considerable variation in $Ev(h)\left(h \in F(f_0^e, f_1^e, g^e)\right)$ in order to apply the method of Section 8.

§12. FURTHER RESULTS AND CONJECTURES

In this section we discuss some, but not all, unsolved problems concerning degrees. We also list some results on degrees obtained since 1963. We regard an unsolved problem as interesting only if it seems likely that its solution requires a new trick. Behind each of our conjectures stand several false but plausible proofs.

We first list some results about recursively enumerable degrees.

(L1) If $\underline{a}$ and $\underline{c}$ are recursively enumerable degrees such that $\underline{a} < \underline{c}$, then there exists a recursively enumerable degree $\underline{b}$ such that $\underline{a} < \underline{b} < \underline{c}$.

(L2) There exist non-recursive, recursively enumerable degrees $\underline{a}$ and $\underline{b}$ such that for no degree $\underline{c}$ is it the case that $\underline{c} \leq \underline{a}$, $\underline{c} \leq \underline{b}$, and $\underline{0} < \underline{c}$.

(L3) There exist incomparable, recursively enumerable degrees $\underline{a}$ and $\underline{b}$ such that $\underline{a}, \underline{b}$ have no greatest lower bound in the degrees or in the recursively enumerable degrees.

(L4) There exist recursively enumerable degrees $\underline{a}$ and $\underline{b}$ such that $\underline{0} < \underline{a} < \underline{b}$ and for no recursively enumerable degree $\underline{c}$ is it the case that $\underline{c} < \underline{b}$ and $\underline{a} \cup \underline{c} = \underline{b}$.

(L5) There exists a recursively enumerable degree $\underline{a}$ such that $\underline{0}^{(n)} < \underline{a}^{(n)} < \underline{0}^{(n+1)}$ for all $n \geq 0$.

(L6) There exists a recursively enumerable degree $\underline{a}$ such that $\underline{a} \cap \underline{b} = \underline{0}$ for no non-recursive, recursively enumerable degree $\underline{b}$.

(L7) There is a non-recursive, recursively enumerable degree $\underline{a} < \underline{0}'$ such that $\underline{a} \cup \underline{b} = \underline{0}'$ for no recursively enumerable degree $\underline{b} < \underline{0}'$.

(L8) There is a degree $< \underline{0}'$ which is incomparable with every recursively enumerable degree other than $\underline{0}$ and $\underline{0}'$.

The results listed above, with the exception of (L5), appear to make necessary use of what might be called the infinite-injury method, an extension of the Friedberg-Muchnik priority method, which was introduced in [21] in order to determine the range of the jump operator restricted to recursively enumerable degrees. On pages 85-86 of Section 6, we compare a priority argument permitting only finitely many injuries to each requirement with one permitting infinitely many injuries to each requirement. If one thinks of Friedberg-Muchnik arguments as making use of limits of convergent sequences of integers, then one can think of infinite-injury arguments as making use of limit points of certain non-convergent sequences.

We proved (L1) in [35]. Our argument was an extension of the proof of Theorem 1 of Section 7 of this monograph. Yates [39] gives another proof of (L1) by means of his own version of the infinite-injury method and the fixed-point theorem of Kleene. (L2) was proved independently by Lachlan [29] and Yates [37]. (L3) was obtained independently by Lachlan [29] and Yates. The ideas used in proving (L2) and (L3) are, as one might rightfully expect, quite similar. (L4) was proved by Lachlan [30]. Subsequently, a stronger version of (L4), namely (L7), was developed by Yates. The proof of (L7) is almost too difficult for even the greatest lover of degrees to endure.

(L5) is a surprising result independently obtained by Lachlan [32] and Martin [33]. Lachlan mixes the proof of Theorem 2 of Section 6 with the fixed-point theorem. Martin also uses Theorem 2 of Section 6, but in a very strange way that is suggestive of the Lowenheim-Skolem theorem.

(L6) appears in Yates [37], and (L8) in Yates [38].

The upper semi-lattice of recursively enumerable degrees is still, in the main, a mystery. We guess there is some simple way of characterizing its ordering, but we are unable to frame a strong conjecture. (L1) suggests that its ordering is homogeneous, but (L2) and (L4) say otherwise. We conjecture only:

(C1) The elementary theory of the ordering of recursively enumerable degrees is decideable;

(C2) For each degree $\underline{d}$, the ordering of degrees recursively enumerable in $\underline{d}$ and $\geq \underline{d}$ is order-isomorphic to the recursively enumerable degrees.

Our next conjecture, suggested by H. Rogers, is:

(C3) For each degree $\underline{d}$, the ordering of degrees $\geq \underline{d}$ is order-isomorphic to the ordering of degrees.

A related conjecture is:

(C4) A partially ordered set P is imbeddable in the degrees if and only if P has cardinality at most that of the continuum and each member of P has at most countably many predecessors.

We know (Section 3) that P is imbeddable in the degrees if each member of P has at most aleph-one successors and at most countably many predecessors, and if P has cardinality at most that of the continuum. Let us see where the arguments of Section 3 break down if used in an attempt to prove (C4). Let b, c, d, and $g \in P$ such that $b \nleq c$, $d \leq c$, and $g \leq c$. Suppose we have imbedded part of P in the degrees so that $\underline{b}$, $\underline{d}$, and $\underline{g}$ have already been chosen. We wish to extend the imbedding to c. The requirements on $\underline{c}$ are: $\underline{b} \nleq \underline{c}$, $\underline{d} \leq \underline{c}$ and $\underline{g} \leq \underline{c}$. Thus we must have $\underline{d} \cup \underline{g} \leq \underline{c}$ and $\underline{b} \nleq \underline{d} \cup \underline{g}$. Suppose $\underline{b}$ and $\underline{d}$ were chosen before $\underline{g}$ was. Then $\underline{g}$ had to be chosen so that $\underline{b} \nleq \underline{d} \cup \underline{g}$. More generally, we were given $\{\underline{b}_i \nleq \underline{d}_i \mid i \in I\}$, where I had cardinality less than that of the continuum, and we had to choose $\underline{g}$ so that $\underline{b}_i \nleq \underline{d}_i \cup \underline{g}$ for all $i \in I$. If I is countable, then the choice of $\underline{g}$ is simple by means of the arguments of Sections 2 and 3. Suppose I is uncountable. Then the choice of $\underline{g}$ is still simple if each member of P has at most aleph-one successors. But, in general, we do not know how to find $\underline{g}$. We believe in the existence of $\underline{g}$, and consequently, in the truth of (C4). We express our belief in $\underline{g}$ by means of the next conjecture:

(C5) If S is a set of independent degrees of cardinality less than that of the continuum, then there exists a degree $\underline{g} \notin S$ such that $S \cup \{\underline{g}\}$ is an independent set of degrees.

Our last conjecture is:

(C6) S is a finite, initial segment of degrees if and only if S is order-isomorphic to a finite, initial segment of some upper semi-lattice with a least member.

It appears to follow from (C6) and some work of J. R. Shoenfield that the elementary theory of the ordering of degrees is unsolvable.

We list two results recently obtained by Yates:

(L9) There is a degree $\underline{a}$ such that $\underline{0} < \underline{a} < \underline{0}'$ and $\underline{0} < \underline{b} \leq \underline{a}$ for no minimal degree $\underline{b}$.

(L10) If $\underline{d}$ is any non-recursive, recursively enumerable degree, then there is a minimal degree less than $\underline{d}$.

We conclude with five open questions about which we have strong feelings that are best concealed:

(Q1) Does there exist a Gödel number e such that for all sets A, w_e^A (the e-th set recursively enumerable in A) is of higher degree than A and of lower degree than A', and such that if A and B have the same degree, then w_e^A and w_e^B have the same degree?

(Q2) Is the elementary theory of the ordering of degrees elementarily equivalent to the elementary theory of the ordering of degrees of arithmetical sets?

(Q3) Is there some simple property of complements of recursively enumerable sets (in the style of Post [17]) which implies non-completeness?

(Q4) Does there exist a sequence of simultaneously recursively enumerable degrees which has a recursively enumerable degree as one of its minimal upper bounds?

(Q5) Is there an elementary difference between the ordering of recursively enumerable degrees and the ordering of metarecursively enumerable degrees [28]?

BIBLIOGRAPHY

[1] Richard M. Friedberg, Two recursively enumerable sets of incomparable degrees of unsolvability, Proc. Nat. Acad. Sci. U.S.A., 43 (1957), 236-238.

[2] _____, The fine structure of degrees of unsolvability of recursively enumerable sets, Summaries of Cornell Univ. Summer Inst. for Symb. Log., (1957), 404-406.

[3] _____, Three theorems on recursive enumeration, J. Symb. Log., 23 (1958), 309-316.

[4] _____, A criterion for completeness of degrees of unsolvability, J. Symb. Log., 22 (1957), 159-160.

[5] Paul R. Halmos, Measure Theory, Princeton, 1950.

[6] John L. Kelley, General Topology, New York, 1955.

[7] S. C. Kleene, Introduction to Metamathematics, New York, Toronto, Amsterdam and Gröningen, 1952.

[8] _____, On the forms of the predicates in the theory of constructive ordinals (second paper), Amer. Jour. Math., 77 (1955), 405-428.

[9] S. C. Kleene and Emil L. Post, The upper semi-lattice of degrees of recursive unsolvability, Ann. of Math., 59 (1954), 379-407.

[10] Daniel Lacombe, Sur le semi-réseau constitué par les degrés d'indécidabilité récursive, C. R. Acad. Sci. Paris, 239 (1954), 1108-1109.

[11] _____, Quelques procédés de definition en topologie récursive, in Constructivity in Mathematics, Amsterdam, 1958, 129-158.

[12] A. Mostowski, Über gewisse universelle Relationen, Ann. Soc. Polon. Math., 17 (1938), 117-118.

[13] A. A. Muchnik, Negative answer to the problem of reducibility of the theory of algorithms (in Russian), Dokl. Akad nauk SSSR, 108 (1956), 194-197.

[14] _____, Solution of Post's reduction problem and of certain other problems in the theory of algorithms (in Russian), Trudy Moskov Mat. Obsc., 7 (1958), 391-405.

[15] John Myhill, Category methods in recursion theory, Pacific Jour. Math., 11 (1961), 1479-1486.

[16] _____, Note on degrees of partial functions, Proc. Amer. Math. Soc., 12 (1961), 519-521.

[17] Emil L. Post, Recursively enumerable sets of positive integers and their decision problems, Bull. Amer. Math. Soc., 50 (1944), 284-316.

[18] Hartley Rogers, Jr., Computing degrees of unsolvability, Math. Ann., 138 (1959), 125-140.

[19] Gerald E. Sacks, On suborderings of degrees of recursive unsolvability, Zeitschr. f. math. Logik und Grundlagen d. Math., 7 (1961), 46- 56.

[20] _____, On the degrees less than $\underline{0}'$, Ann. of Math., 77 (1963), 211-231.

[21] _____, Recursive enumerability and the jump operator, Trans. of the Amer. Math. Soc., 108 (1963), 223-239.

[22] J. R. Shoenfield, An uncountable set of incomparable degrees, Proc. Amer. Math. Soc., 11 (1960), 61-62.

[23] _____, On degrees of unsolvability, Ann. of Math., 69 (1959), 644-653.

[24] W. Sierpinski, General Topology, Toronto, 1952.

[25] Clifford Spector, On degrees of recursive unsolvability, Ann. of Math., 64 (1956), 581-592.

[26] _____, Measure-theoretic construction of incomparable hyperdegrees, J. Symb. Log., 23 (1958), 280-288.

[27] D. Titgemeyer, Untersuchungen über die Struktur des Kleene-Postschen Halbverbandes der Grade der rekursiven Unlösbarkeit, Doctoral Thesis, Westfälischen Wilhelmsuniversität zu Munster, 1962.

The following references were added in 1966.

[28] G. Kreisel and G. Sacks, Metarecursive sets, J. Symb. Log., 30(1965) 318-338.

[29] A. H. Lachlan, Lower bounds for pairs of recursively enumerable degrees, Proc. Lond. Math. Soc., to appear.

[30] ______, The impossibility of finding relative complements for recursively enumerable degrees, J. Symb. Log., to appear.

[31] ______, The priority method I, Zeitschr. f. math. Logik und Grundlagen d. Math., to appear.

[32] ______, On a problem of G. E. Sacks, Proc. Amer. Math. Soc., 16 (1965), 972-979.

[33] D. A. Martin, On a question of G. E. Sacks, J. Symb. Log., to appear.

BIBLIOGRAPHY

[34] R. W. Robinson, Ph.D. Thesis, Cornell University, 1966.

[35] G. E. Sacks, The recursively enumerable degrees are dense, Ann. of Math., 80 (1964), 300-312.

[36] _____, Measure-theoretic uniformity, Bull. Amer. Math. Soc., to appear.

[37] C. E. M. Yates, A minimal pair of recursively enumerable degrees, J. Symb. Log., to appear.

[38] _____, Recursively enumerable degrees and the degrees less than 0', Proc. Leicester Logic Conference, North-Holland, to appear.

[39] _____, On the degrees of index sets I, Trans. Amer. Math. Soc., to appear.

Printed in the United States
737500003B

9 780691 079417